Technical Instrumentation Coloring Book For Students

Volume 2

Ezekiel Caudill
and
Fox Howell

Copyright © 2021 PAC Technical Publication Group
All rights reserved.
ISBN: 9798508644949

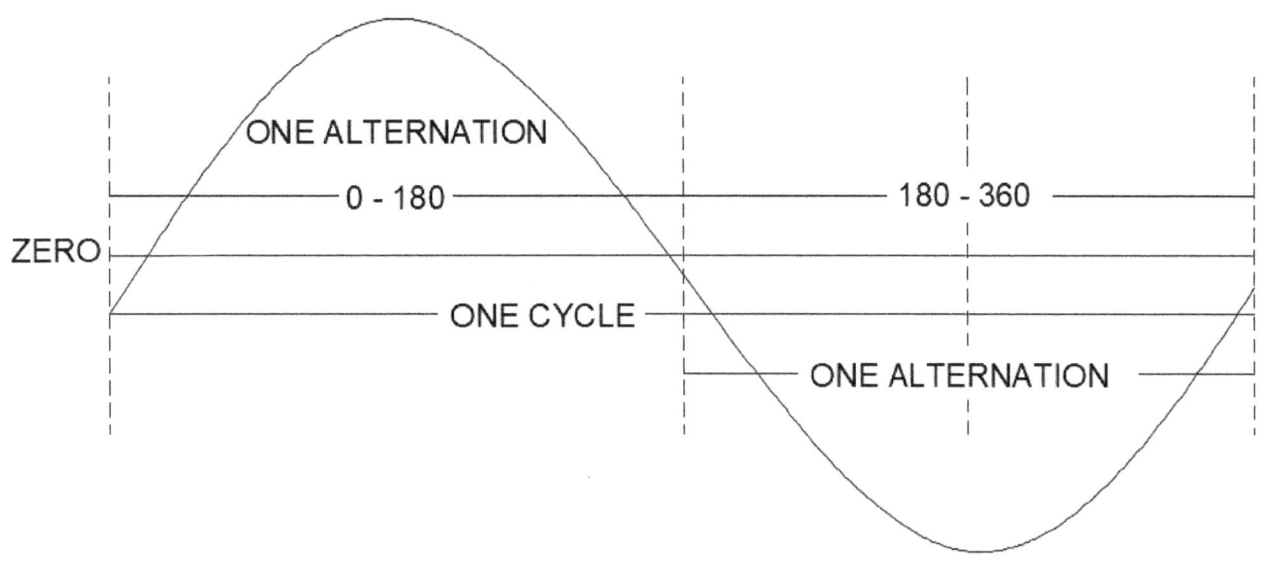

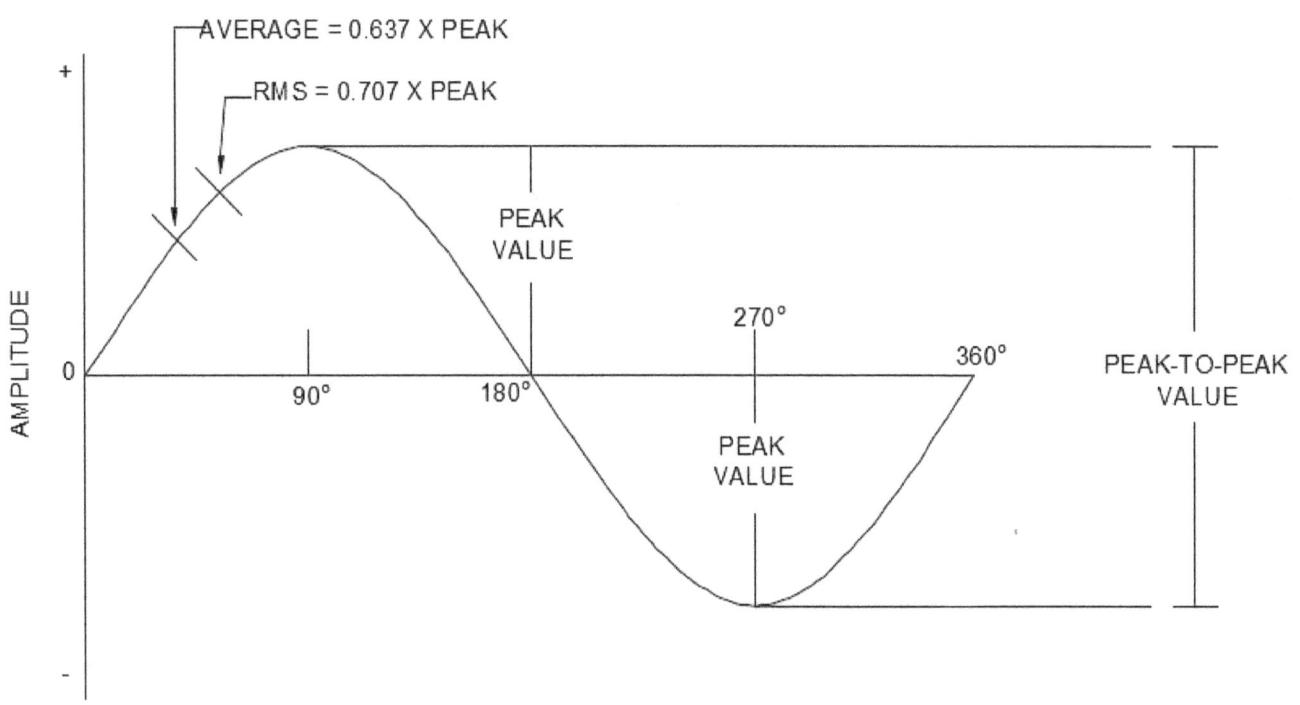

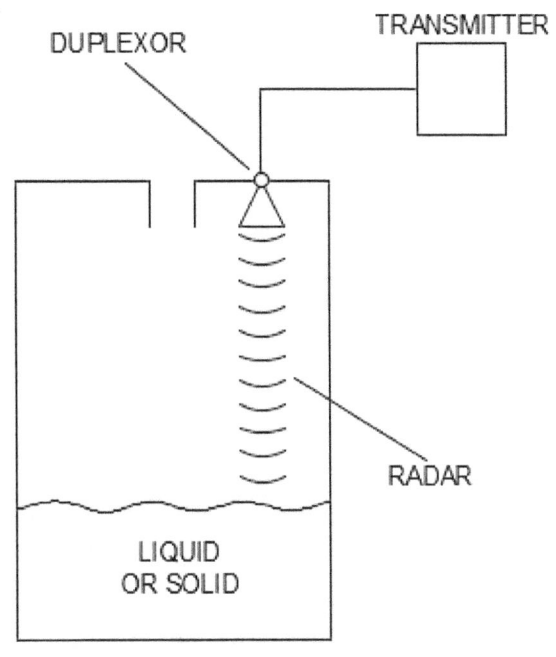

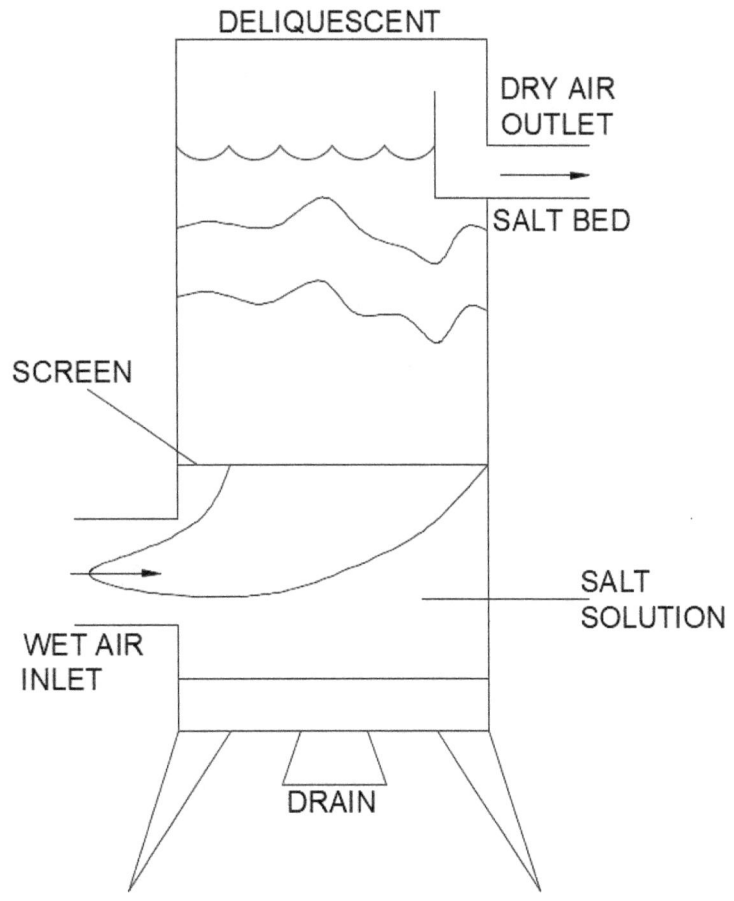

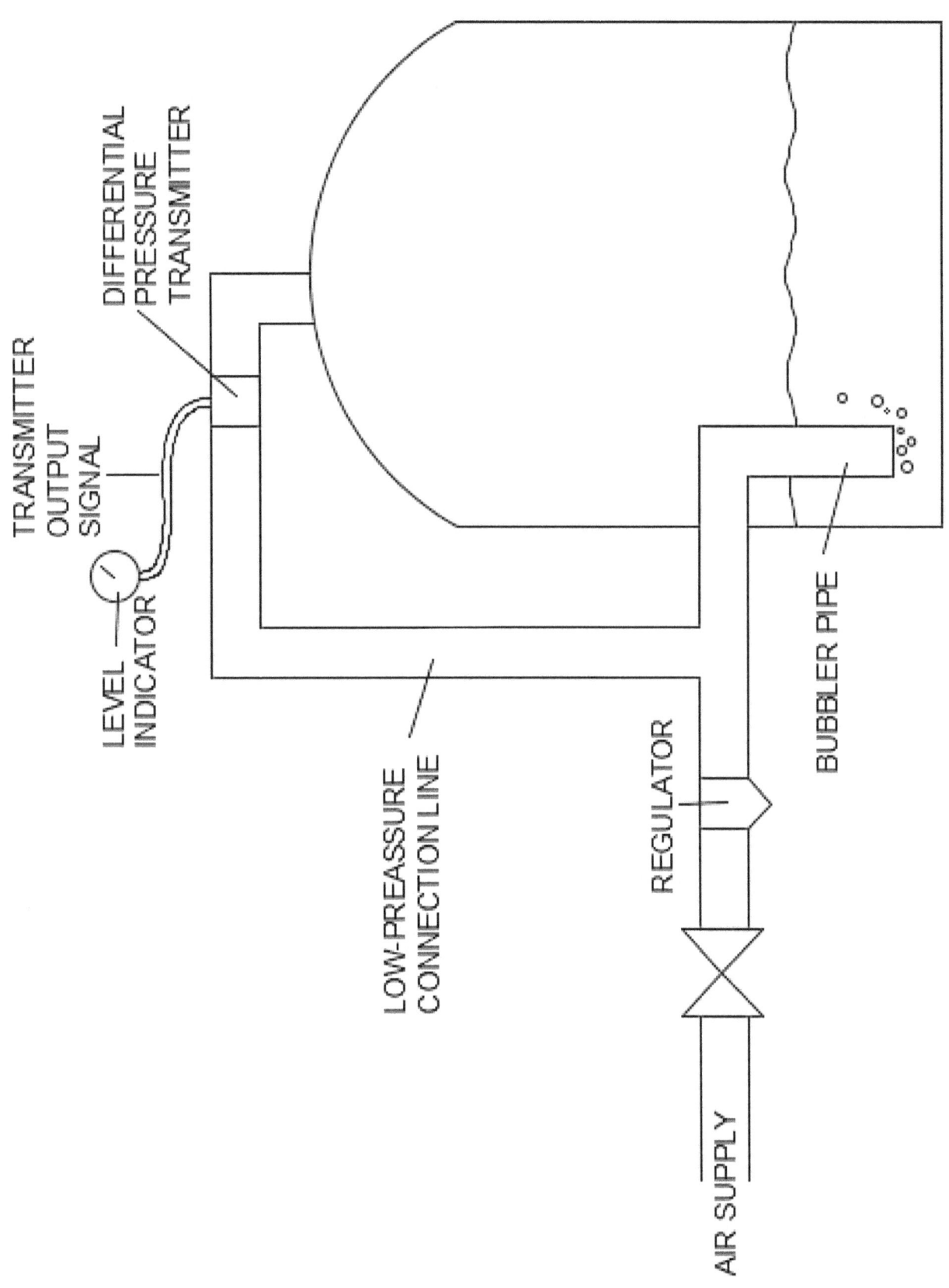

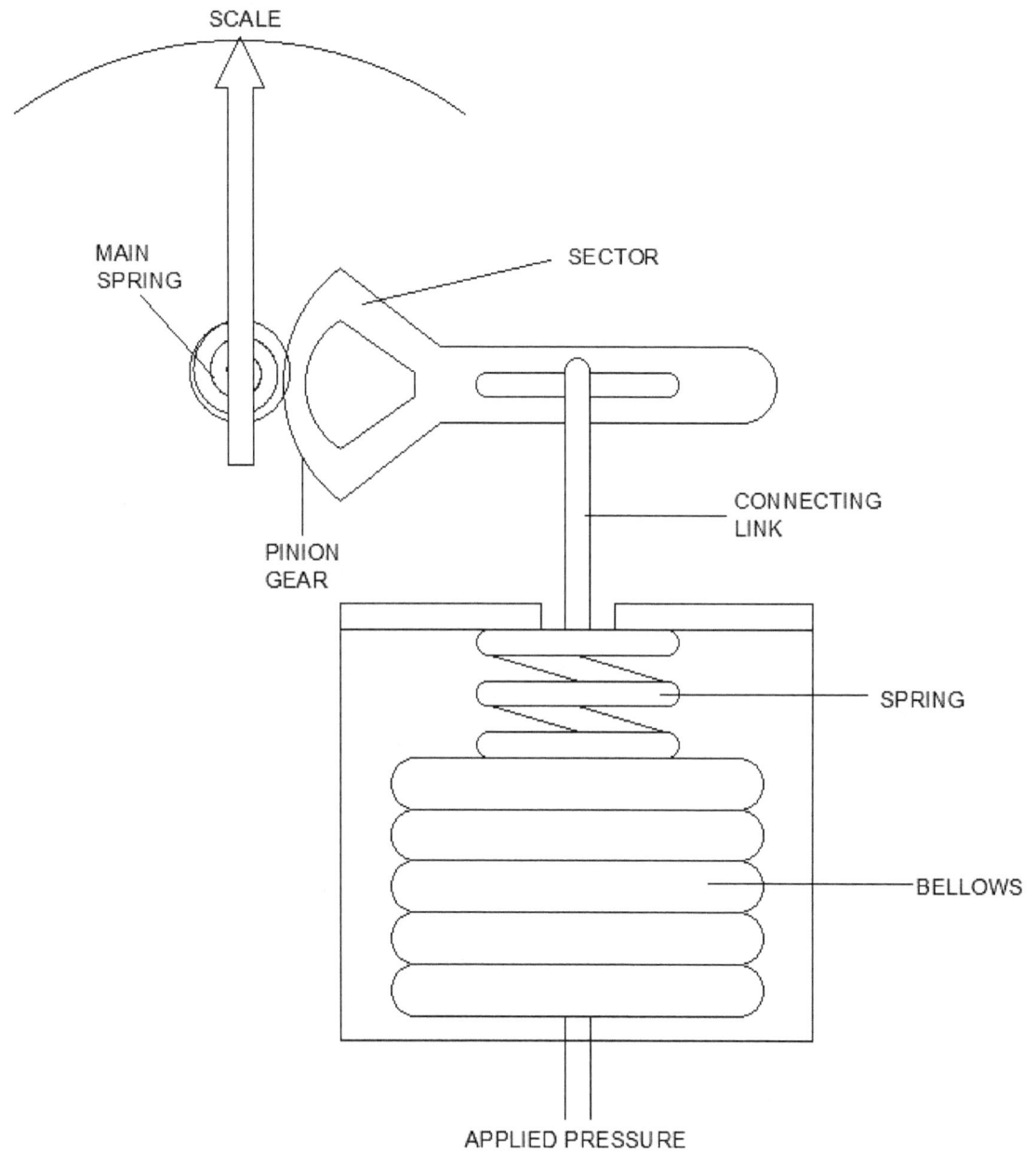

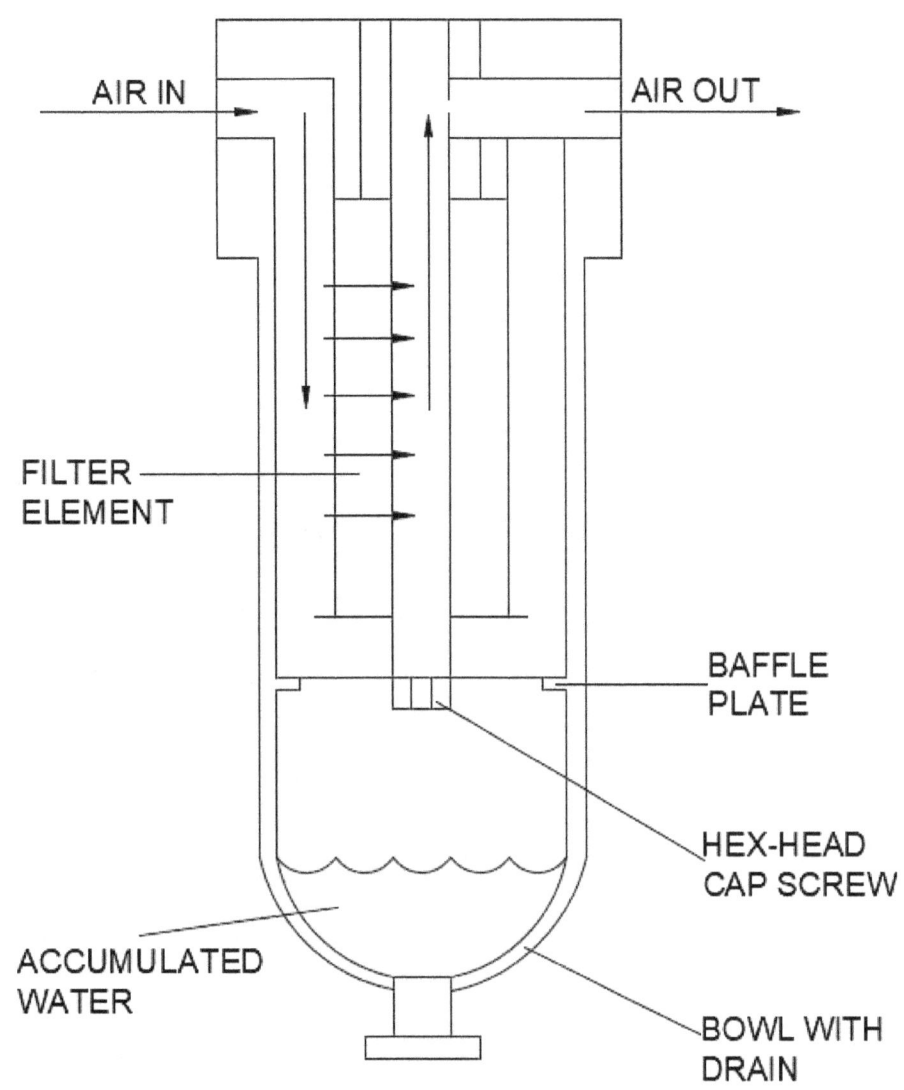

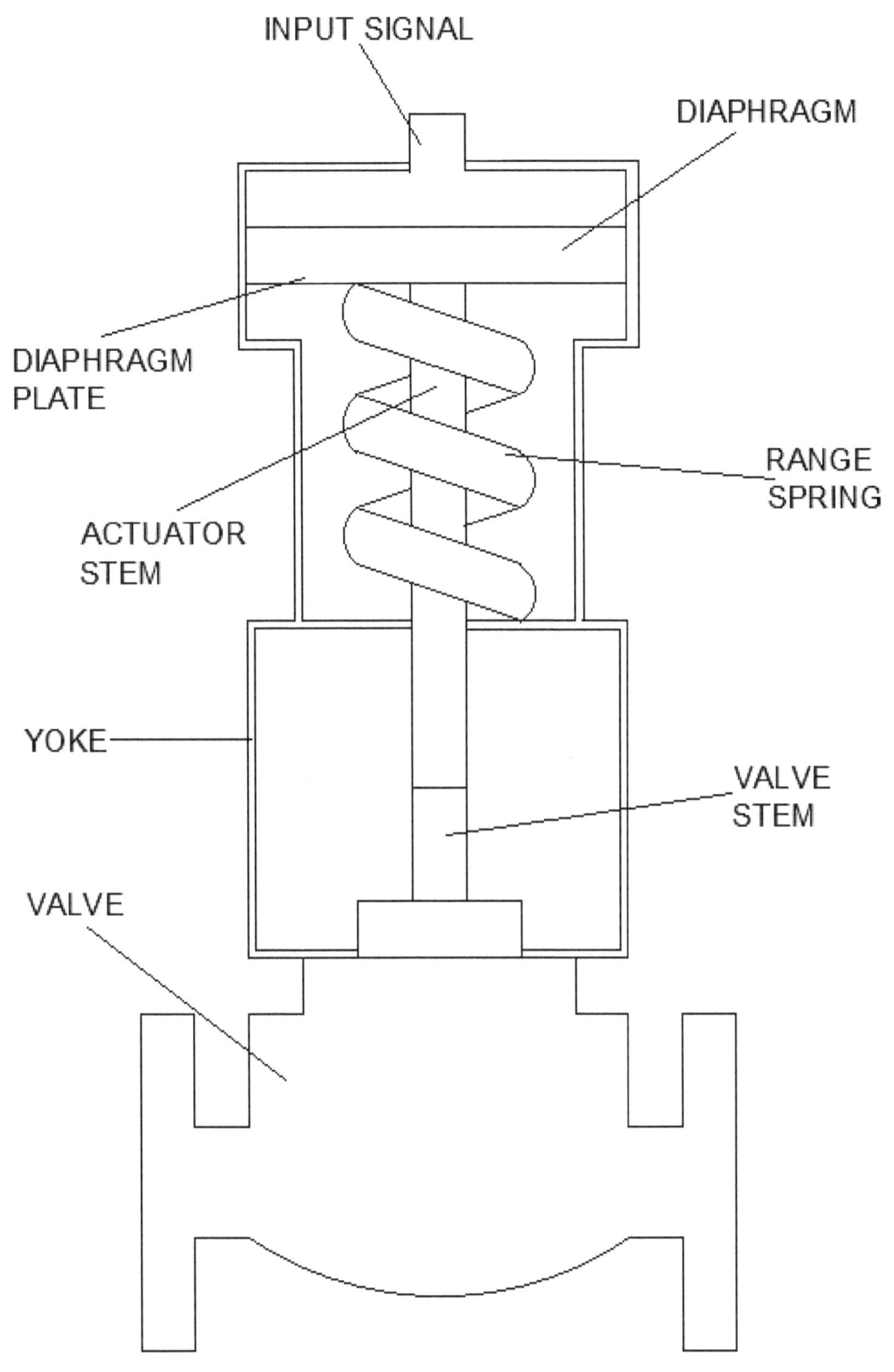

17

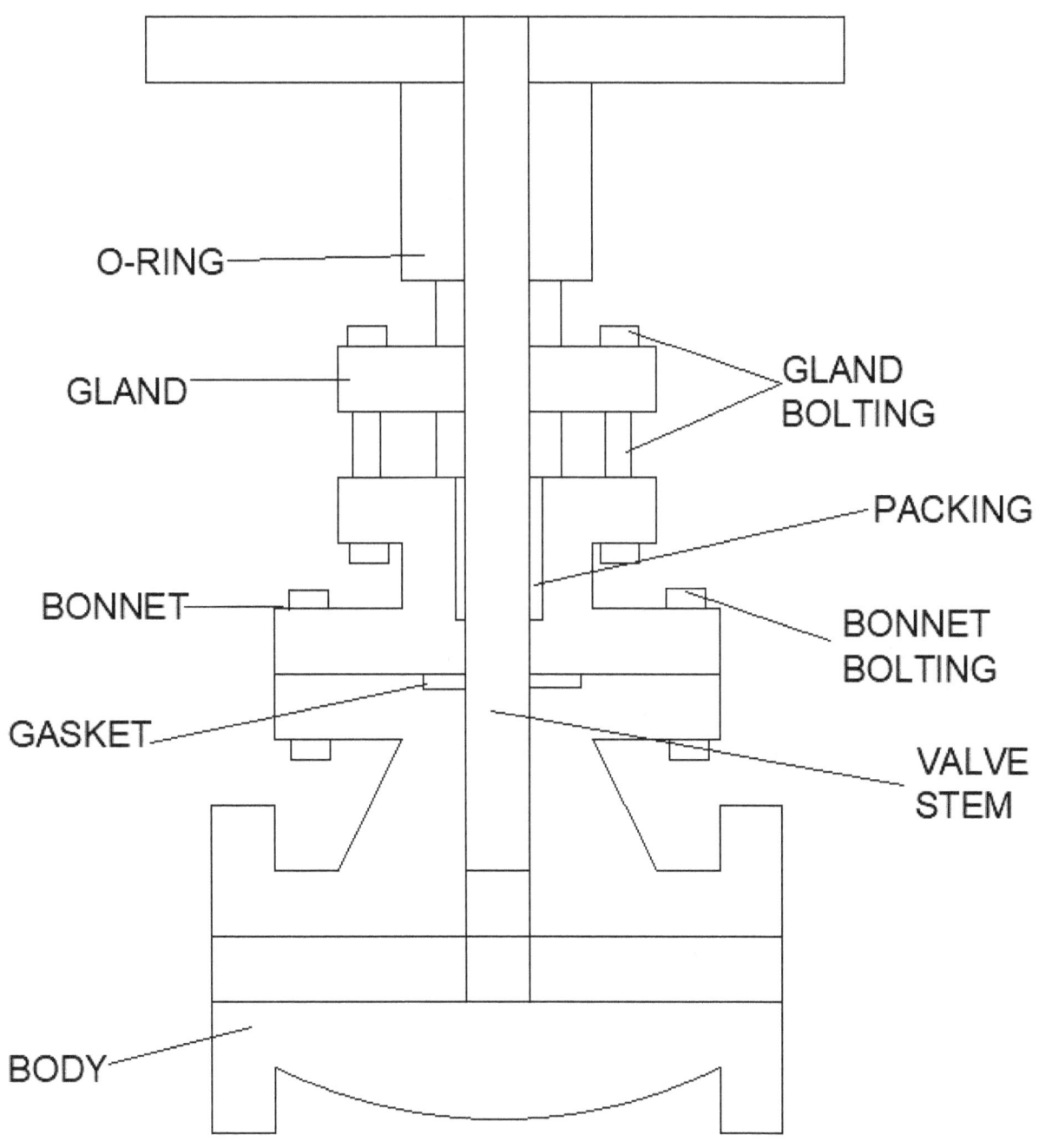

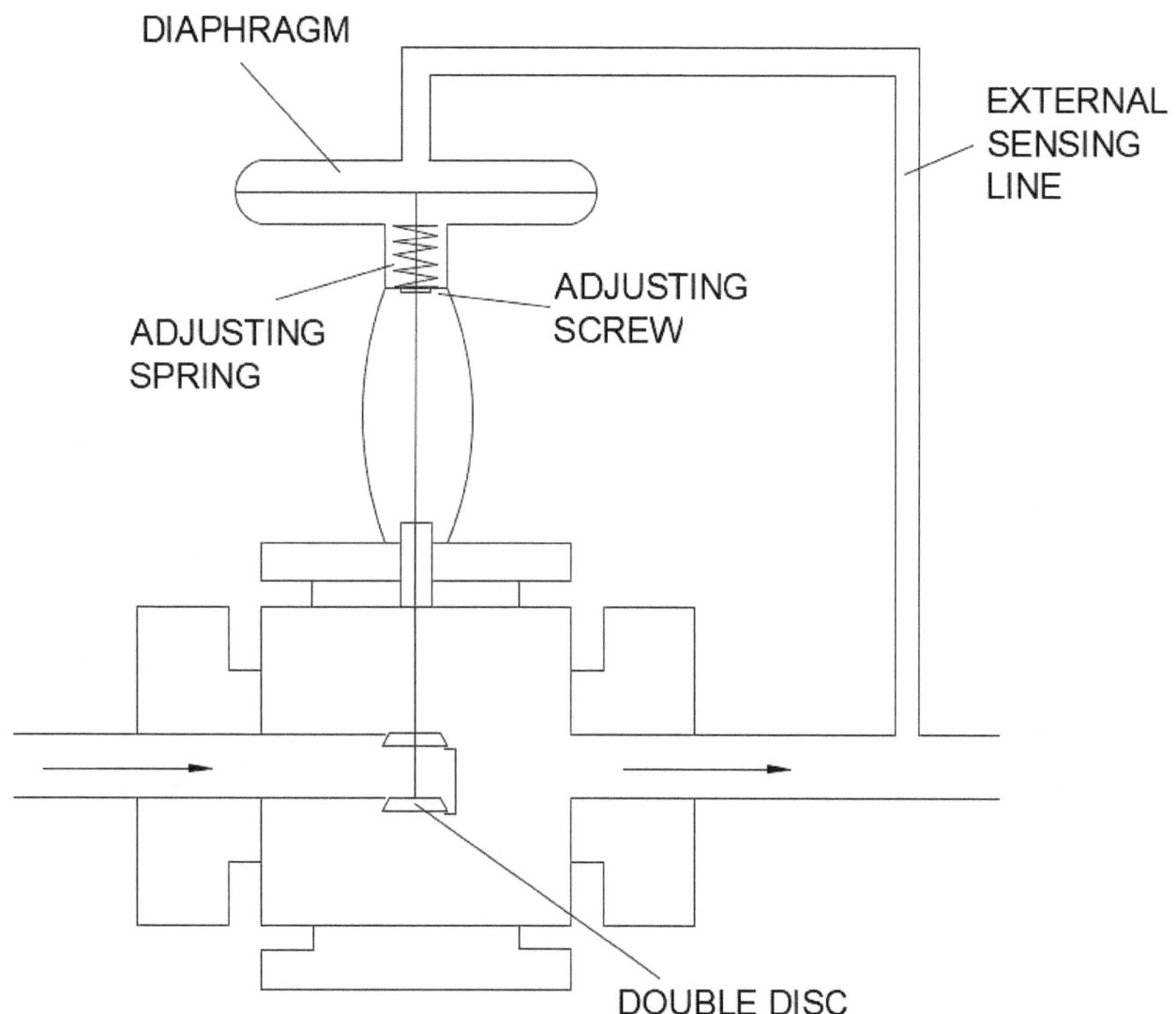

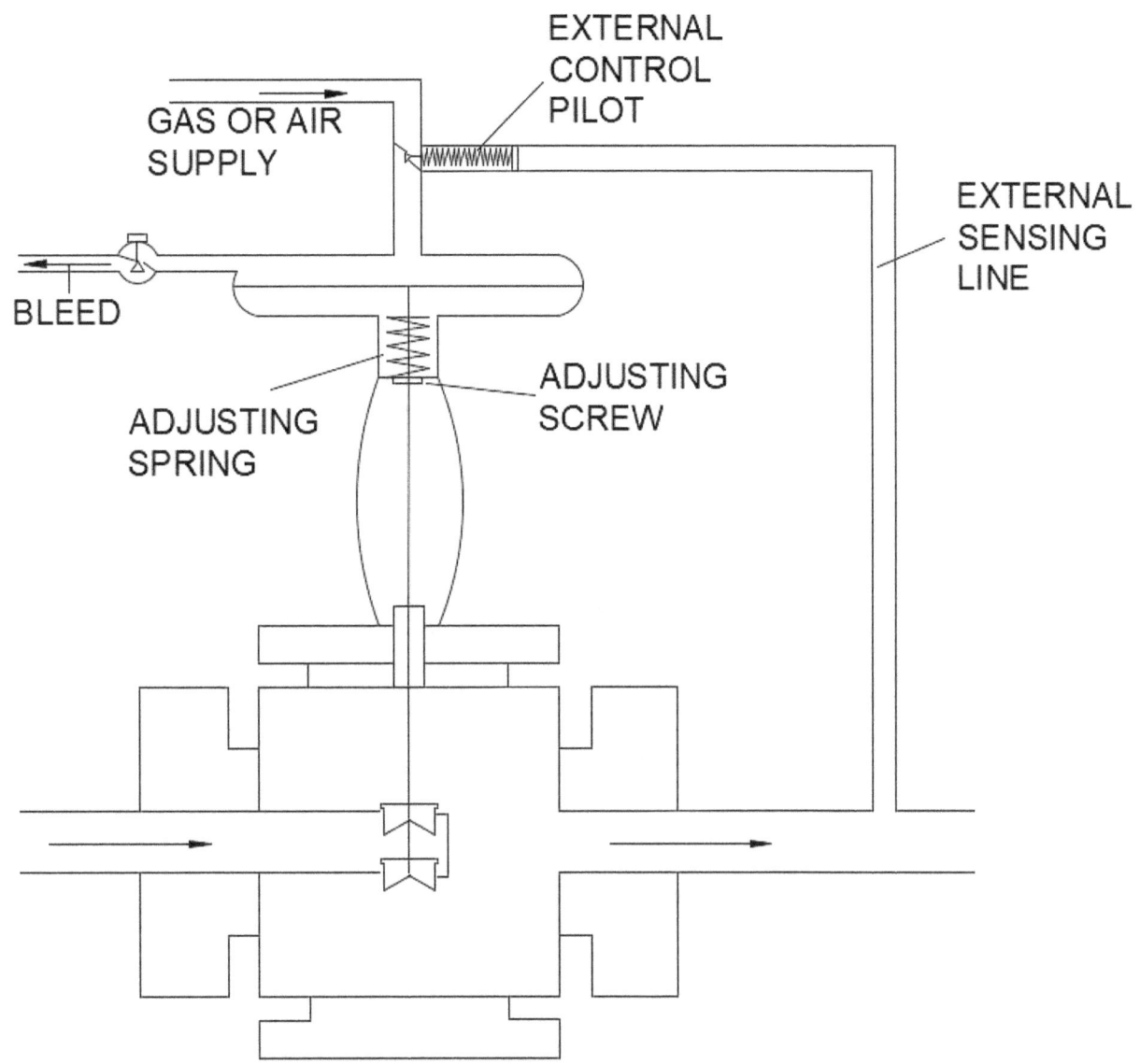

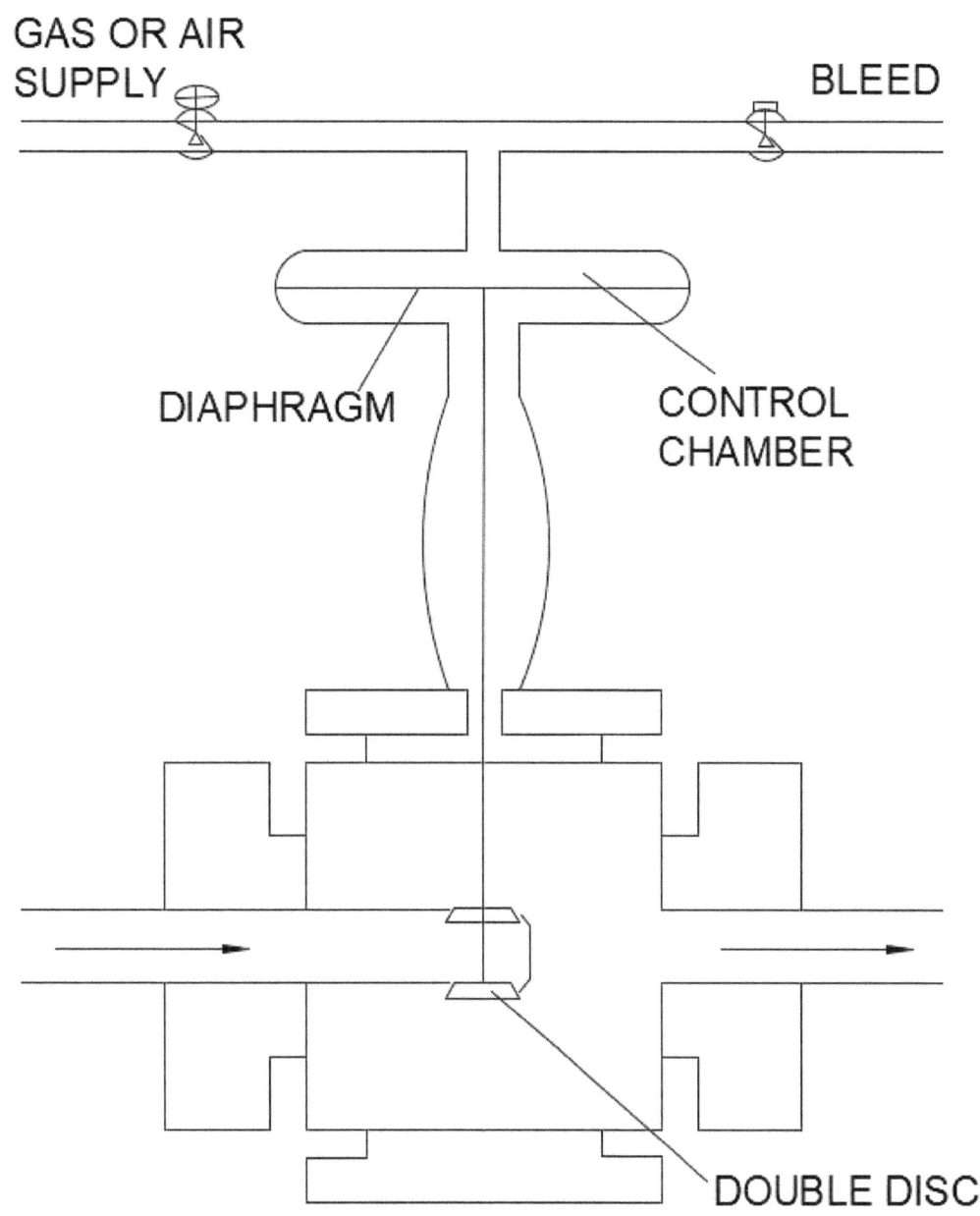

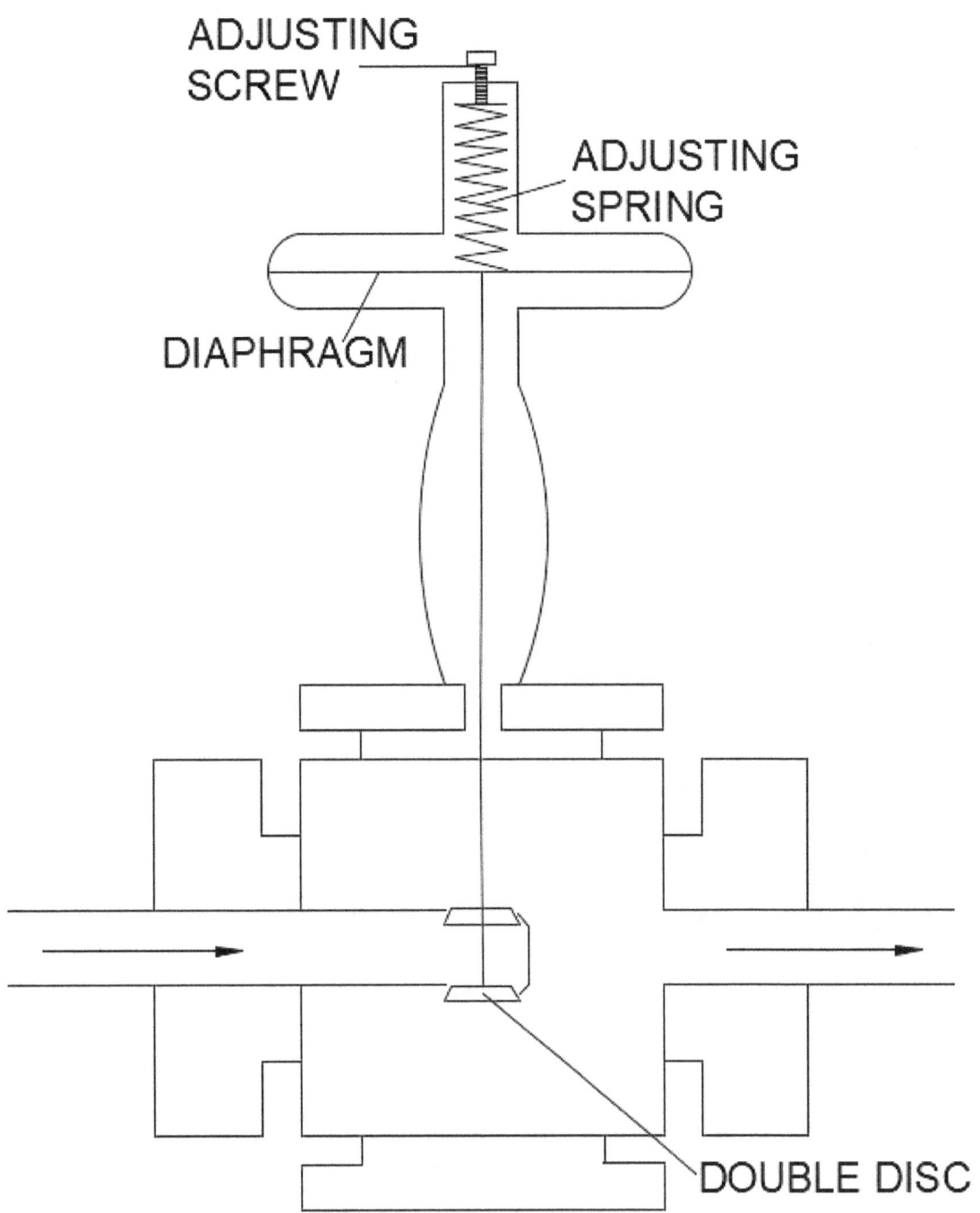

⸻ (thick solid line)	INSTRUMENT SUPPLY
─── (thin solid line)	CONNECTION TO PROCESS
─//─//─//─//─	PNEUMATIC SIGNAL
─ ─ ─ ─ ─ ─	ELECTRICAL SIGNAL
─⌐─⌐─⌐─⌐─	HYDRAULIC SIGNAL
─×─×─×─×─	CAPILLARY TUBE
∿∿∿∿	ELECTROMAGNETIC OR SONIC SIGNAL
─○─○─○─○─	SOFTWARE LINK
─●─●─●─●─	MECHANICAL CONNECTION

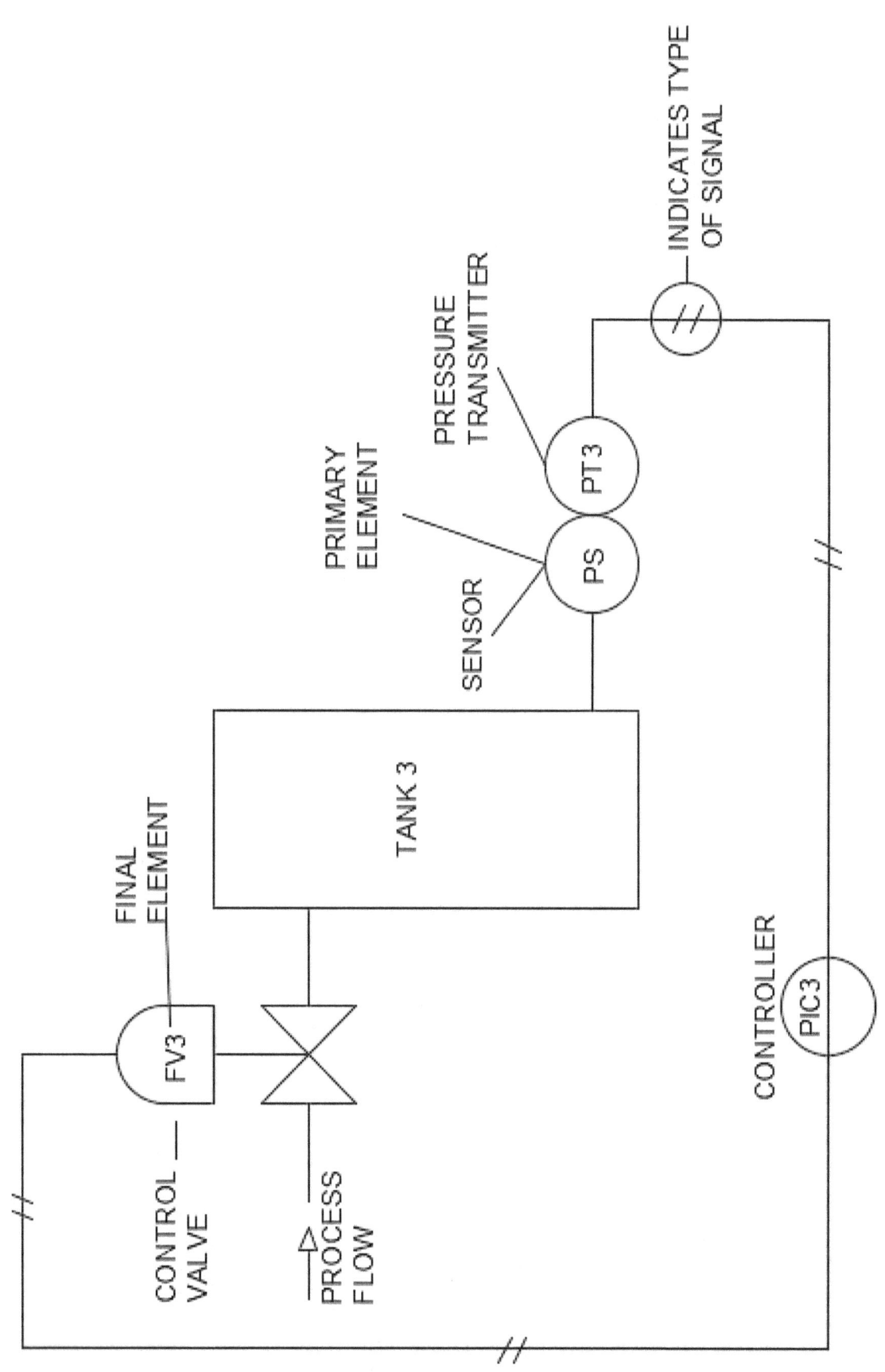

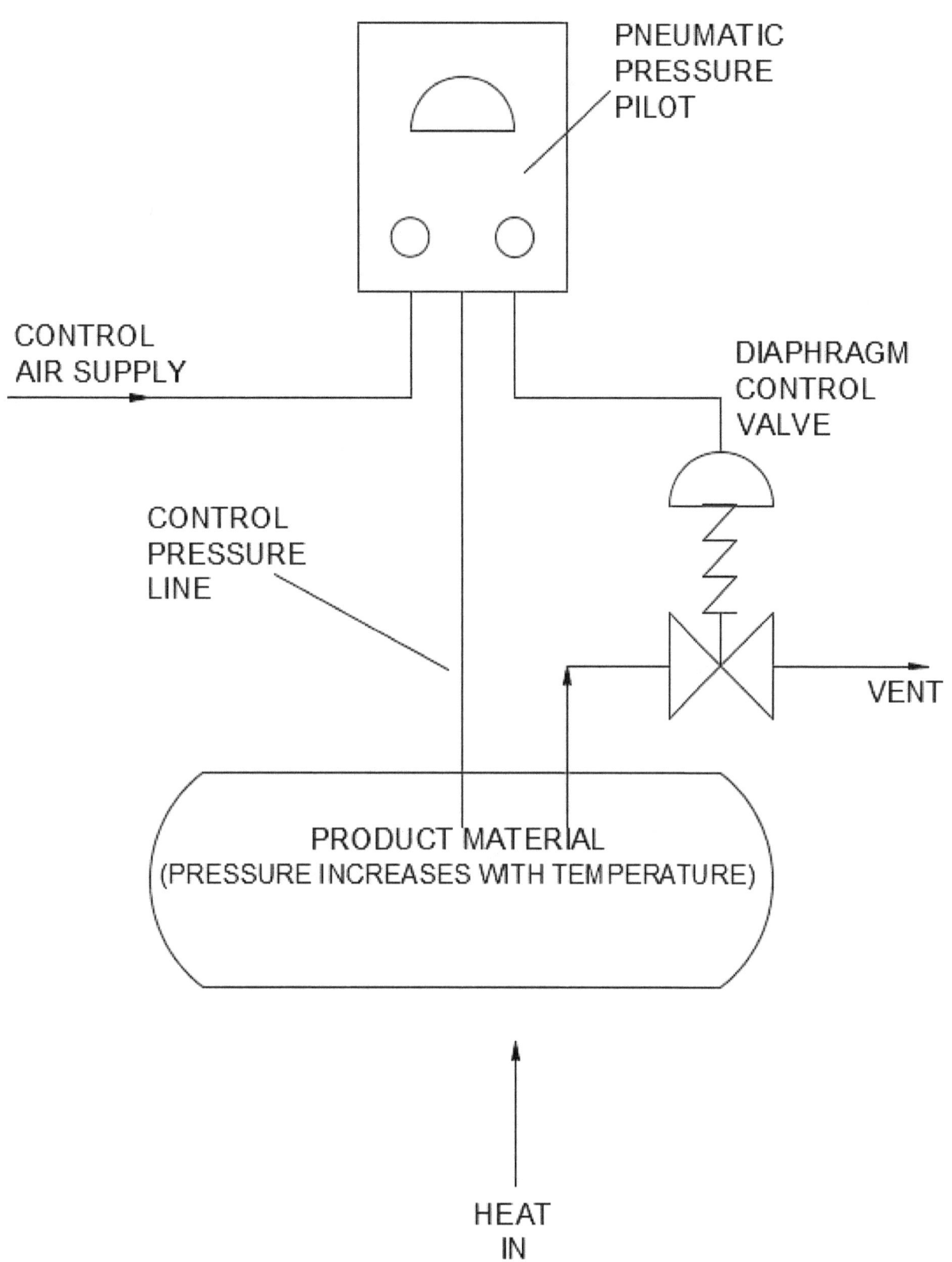

"My brain is only a receiver,
in the Universe there is a core
from which we obtain knowledge,
strength and inspiration."

- *Nikola Tesla*

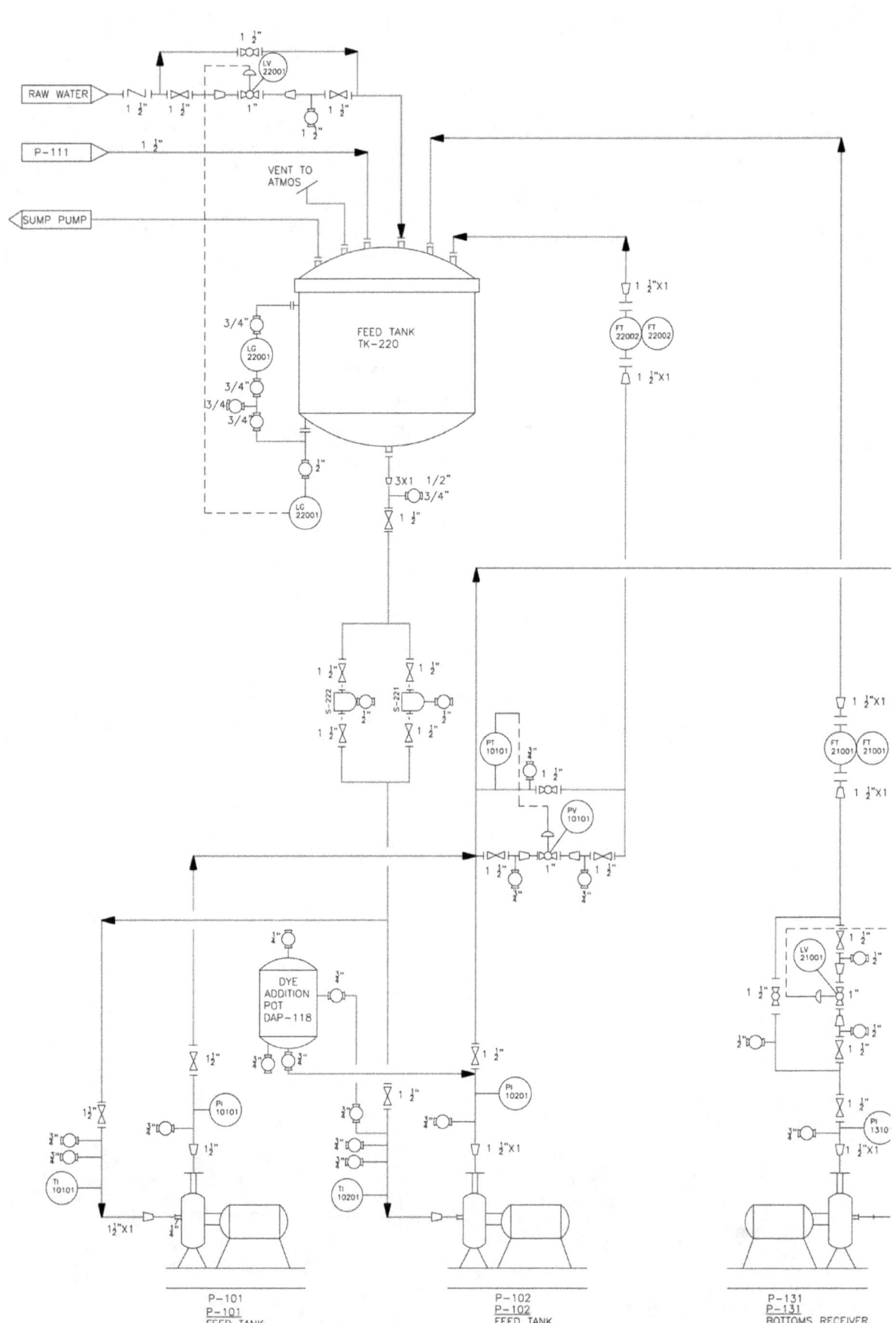

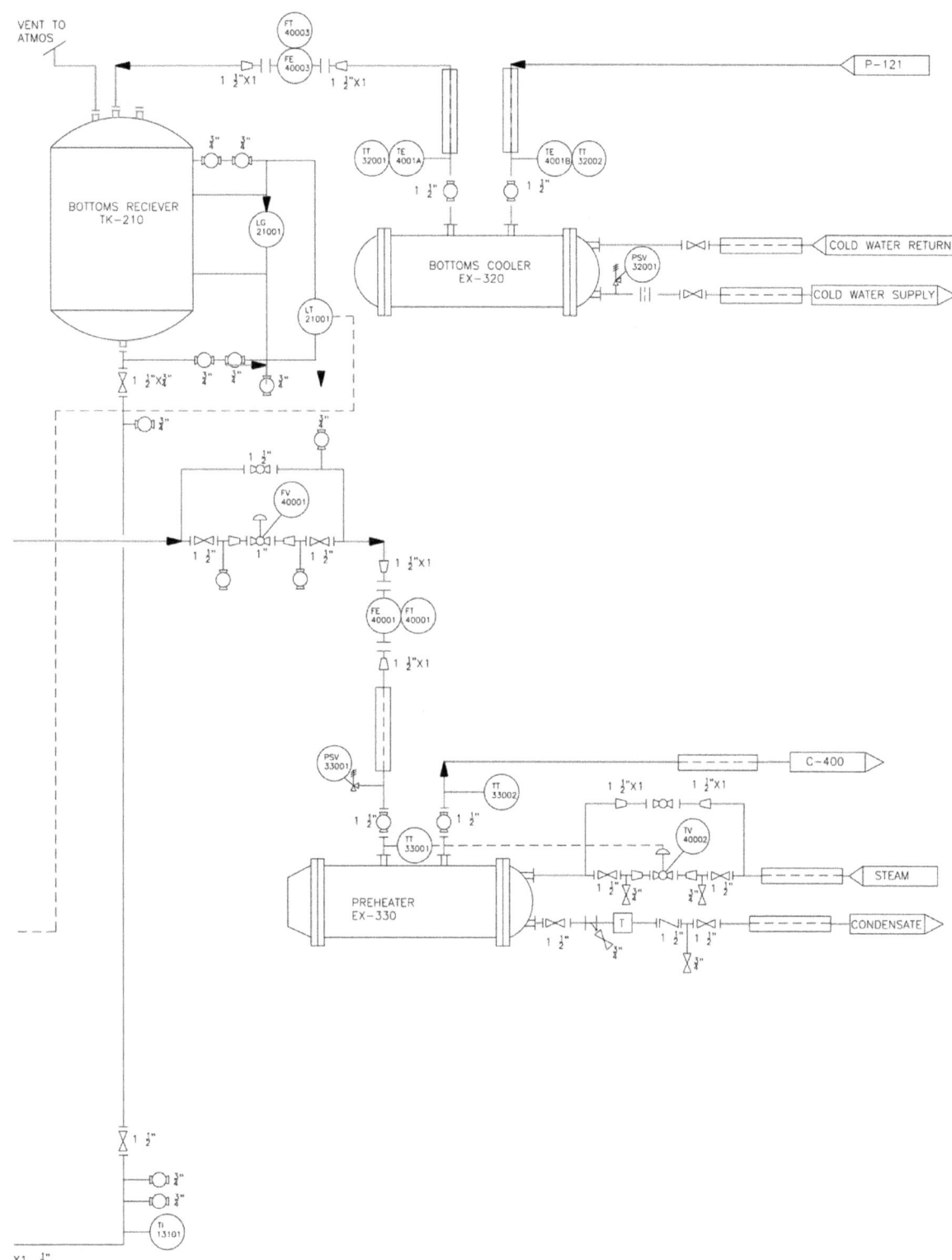

"Would I rather be feared or loved? Easy. Both.
I want people to be afraid of how much they love me."

- *Michael Scott*

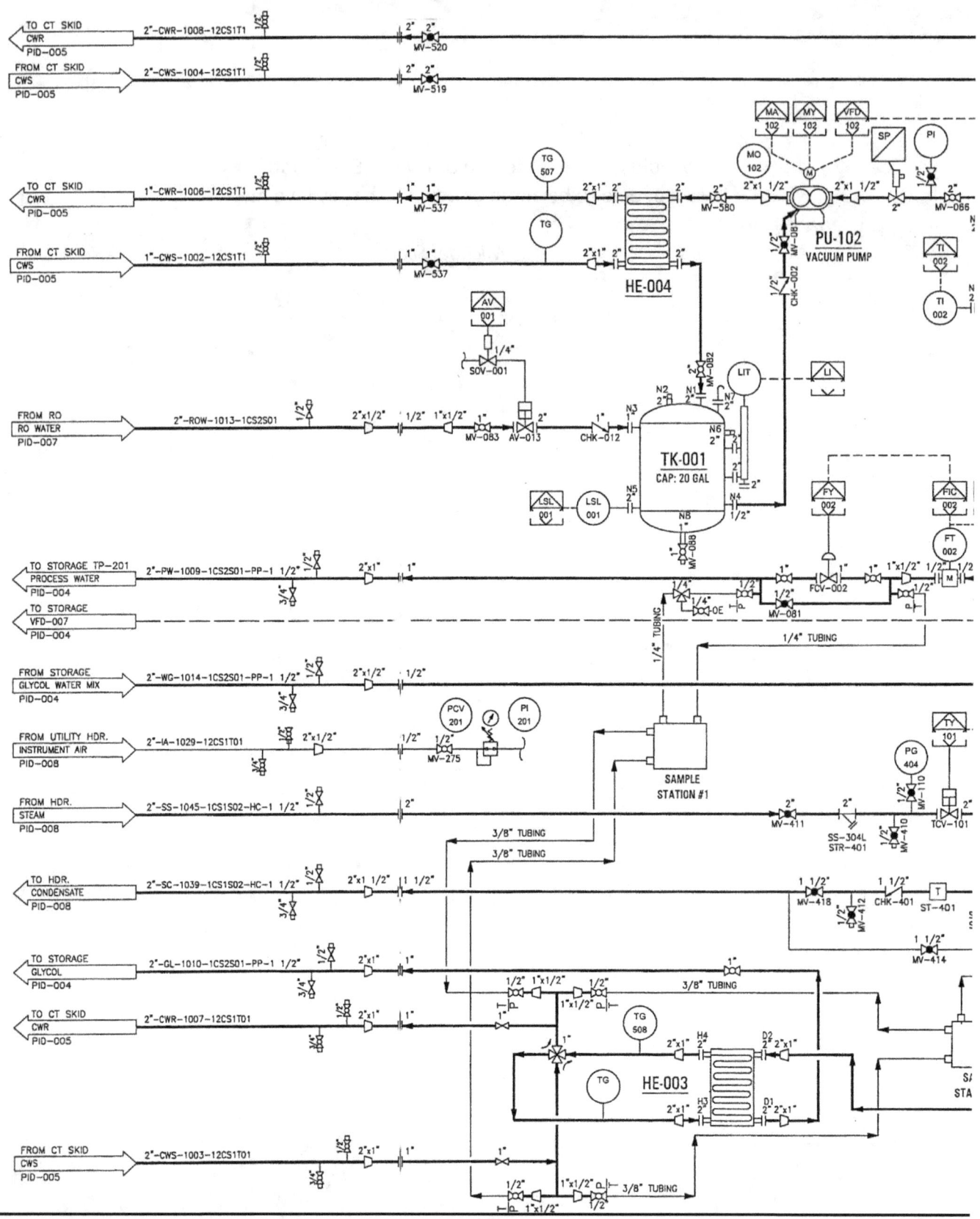

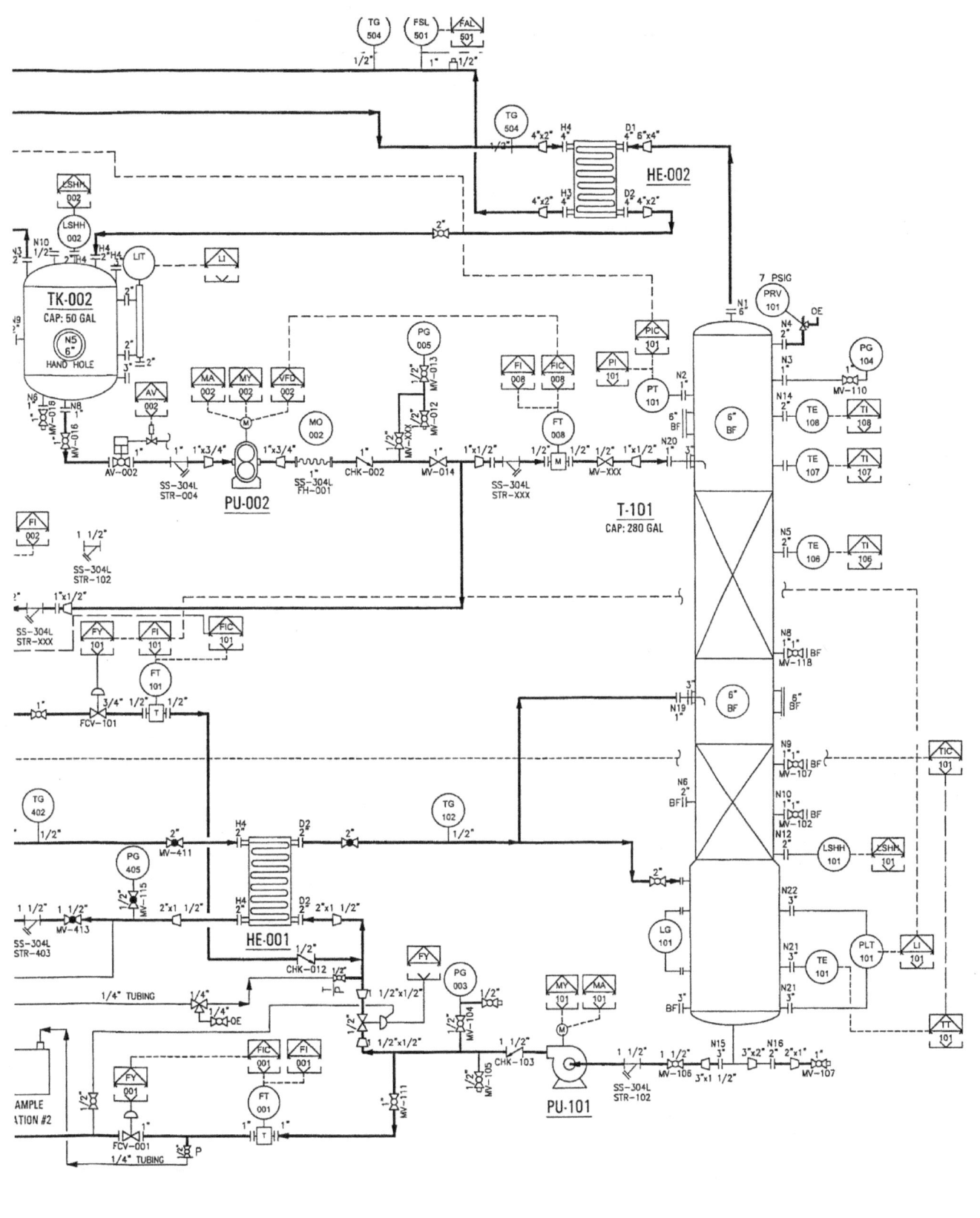

"Do you want ants?
Because that's how you get ants."

- *Sterling Archer*

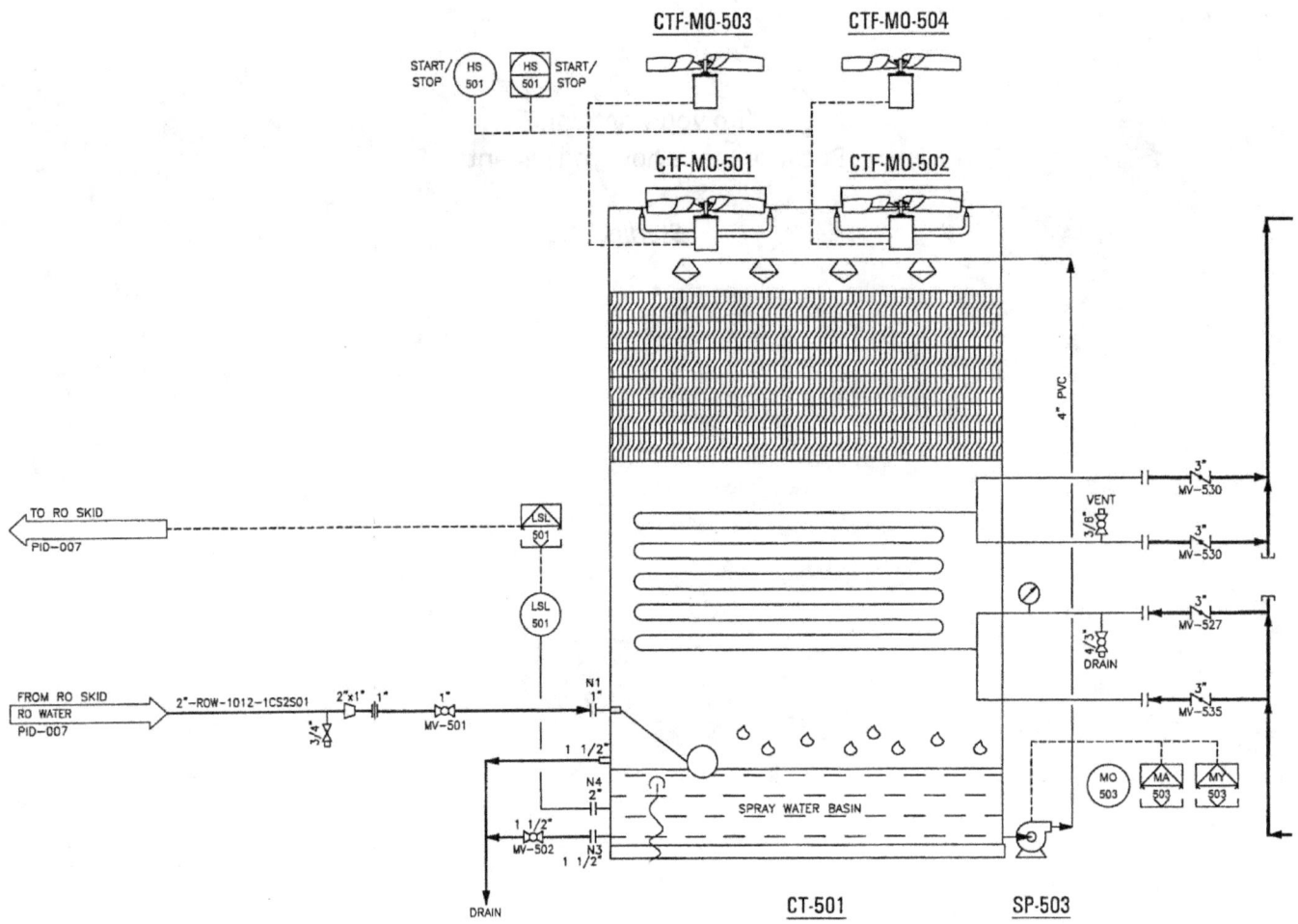

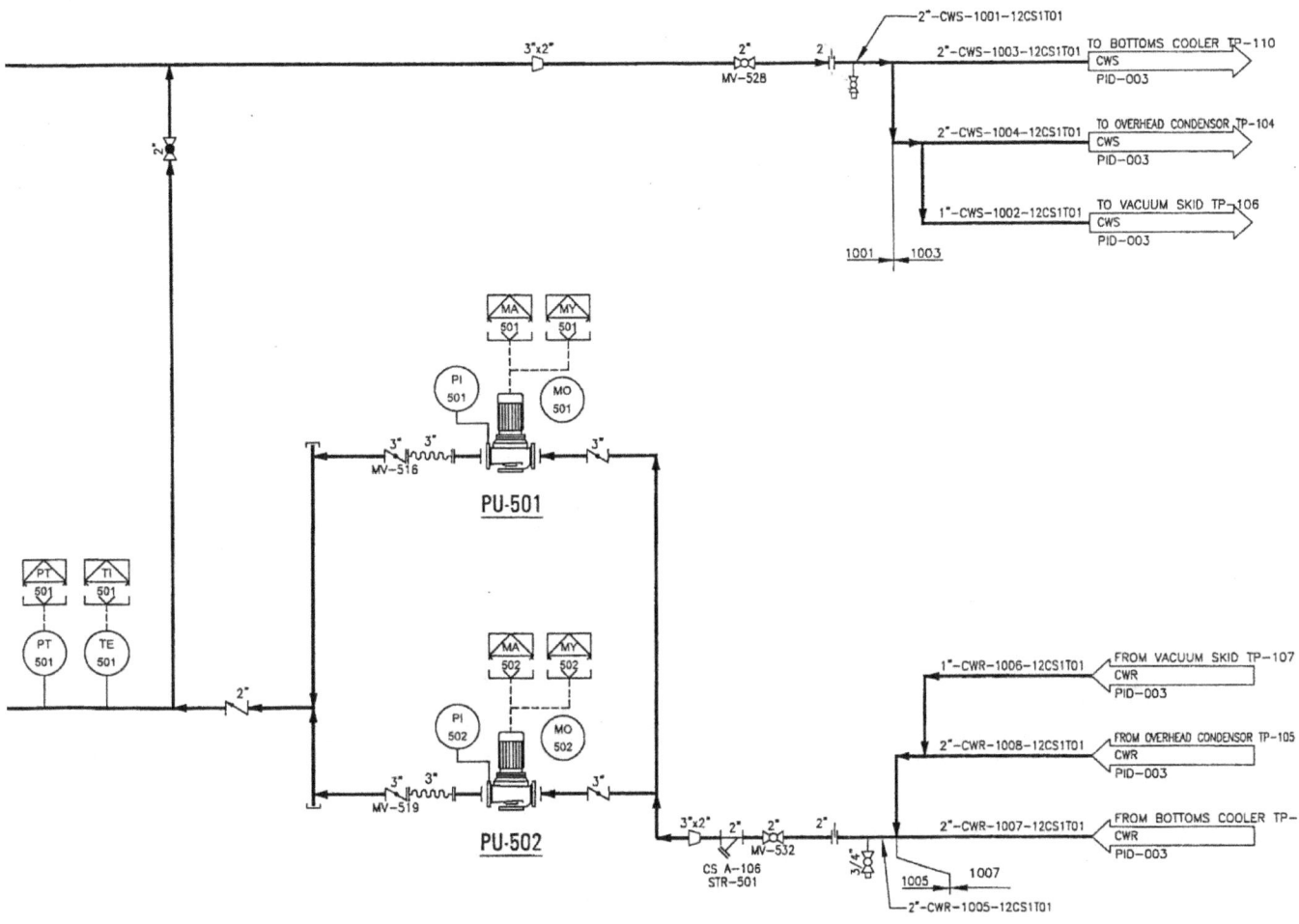

"I know that new situations can be intimidating.
You're lookin' around and it's all scary and different,
but y'know...meeting them head-on, charging
into 'em like a bull – that's how we grow as people."

- *Rick Sanchez*

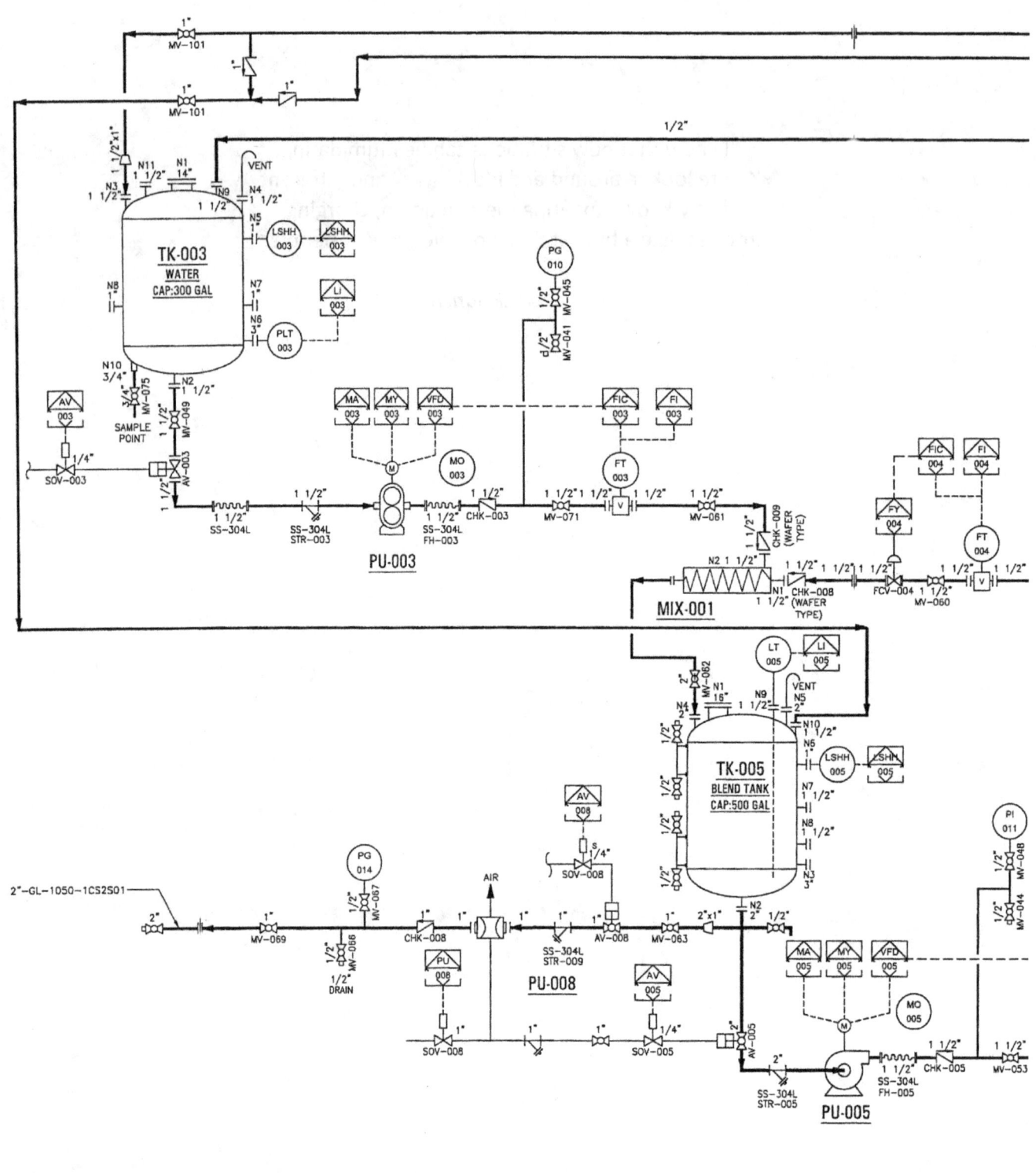

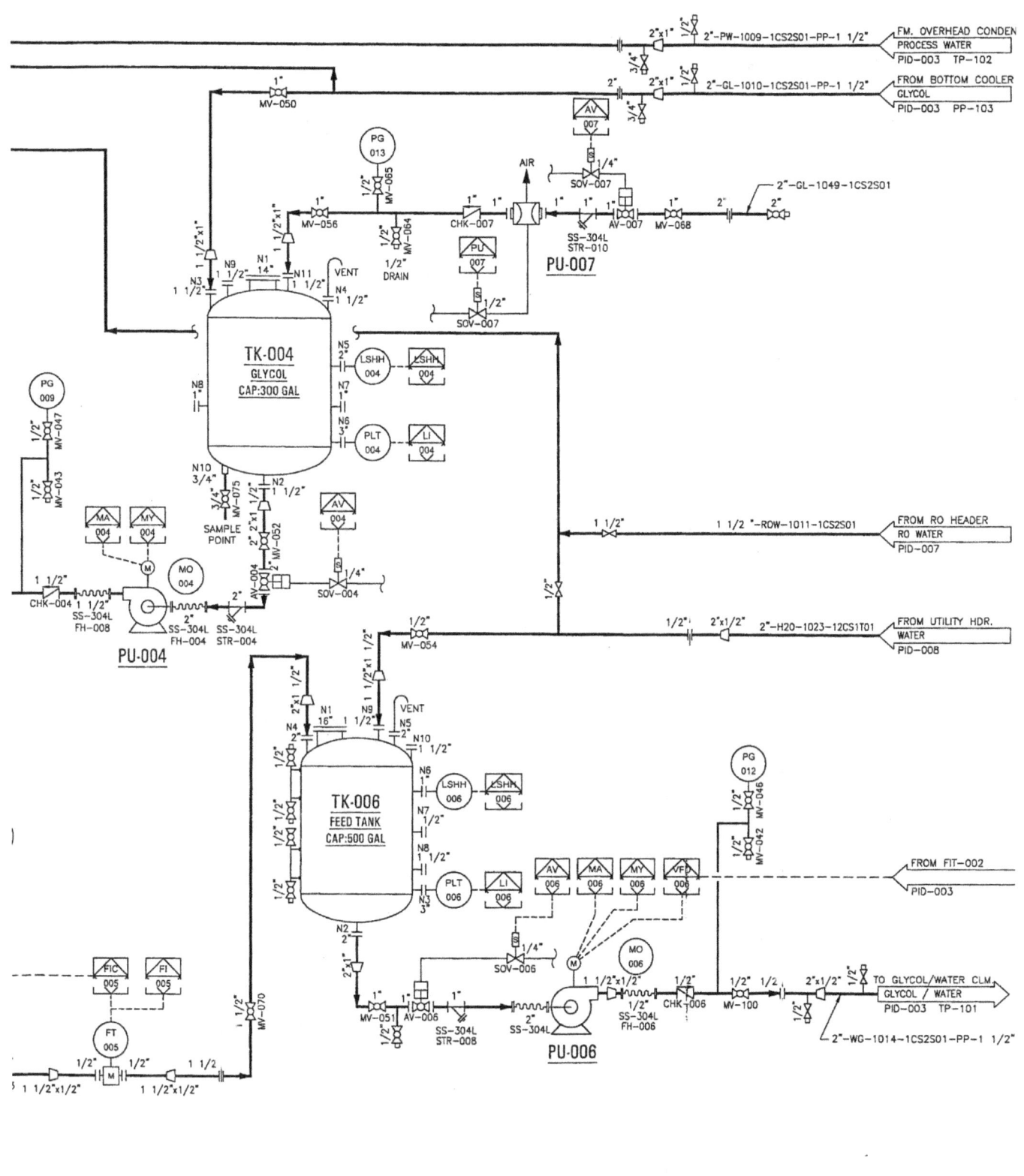

"YOU ARE ALL BETTER THAN YOU THINK YOU ARE,
YOU ARE JUST DESIGNED NOT TO BELIEVE IT
WHEN YOU HEAR IT FROM YOURSELF."

- *JEFF WINGER*

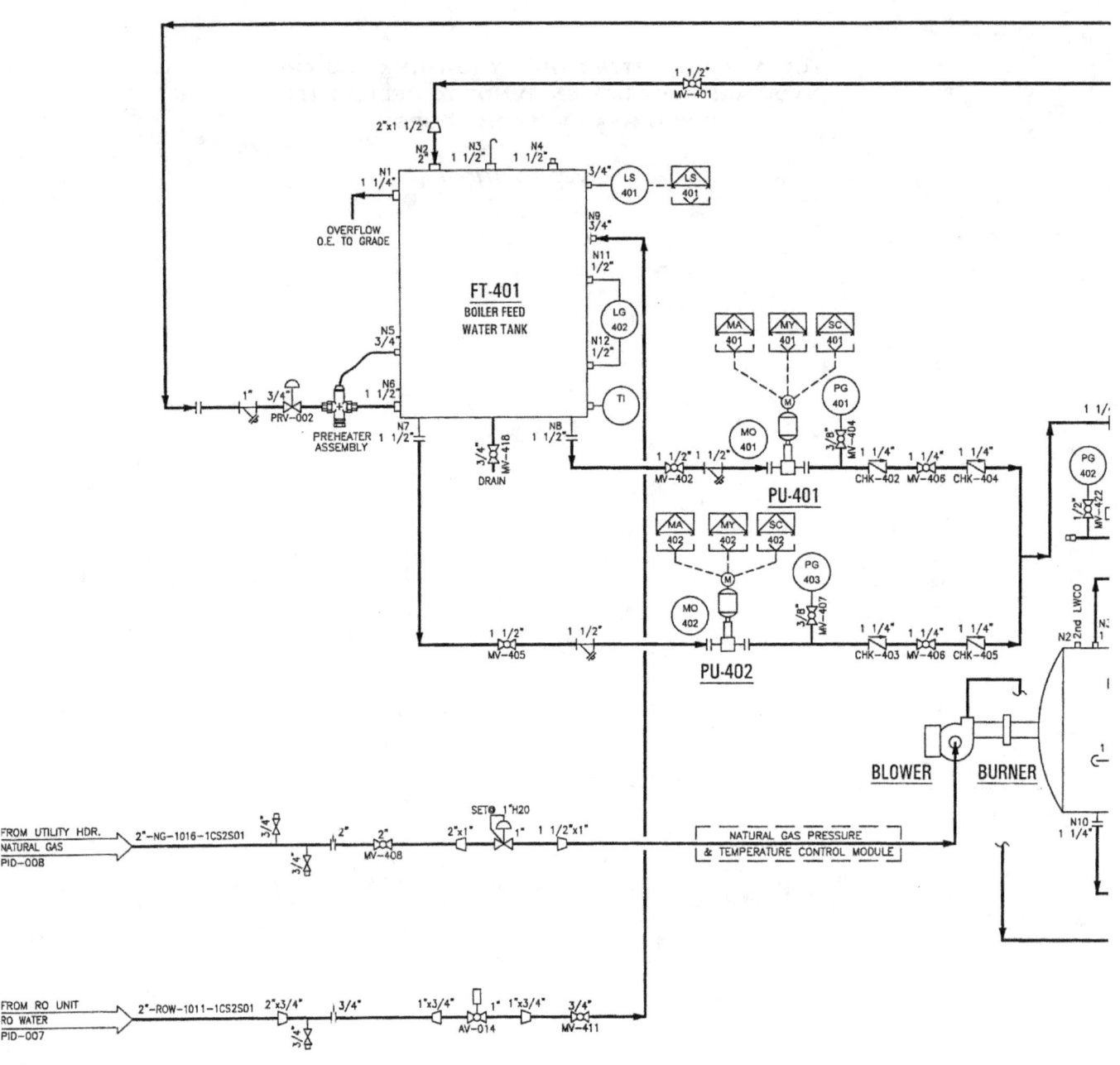

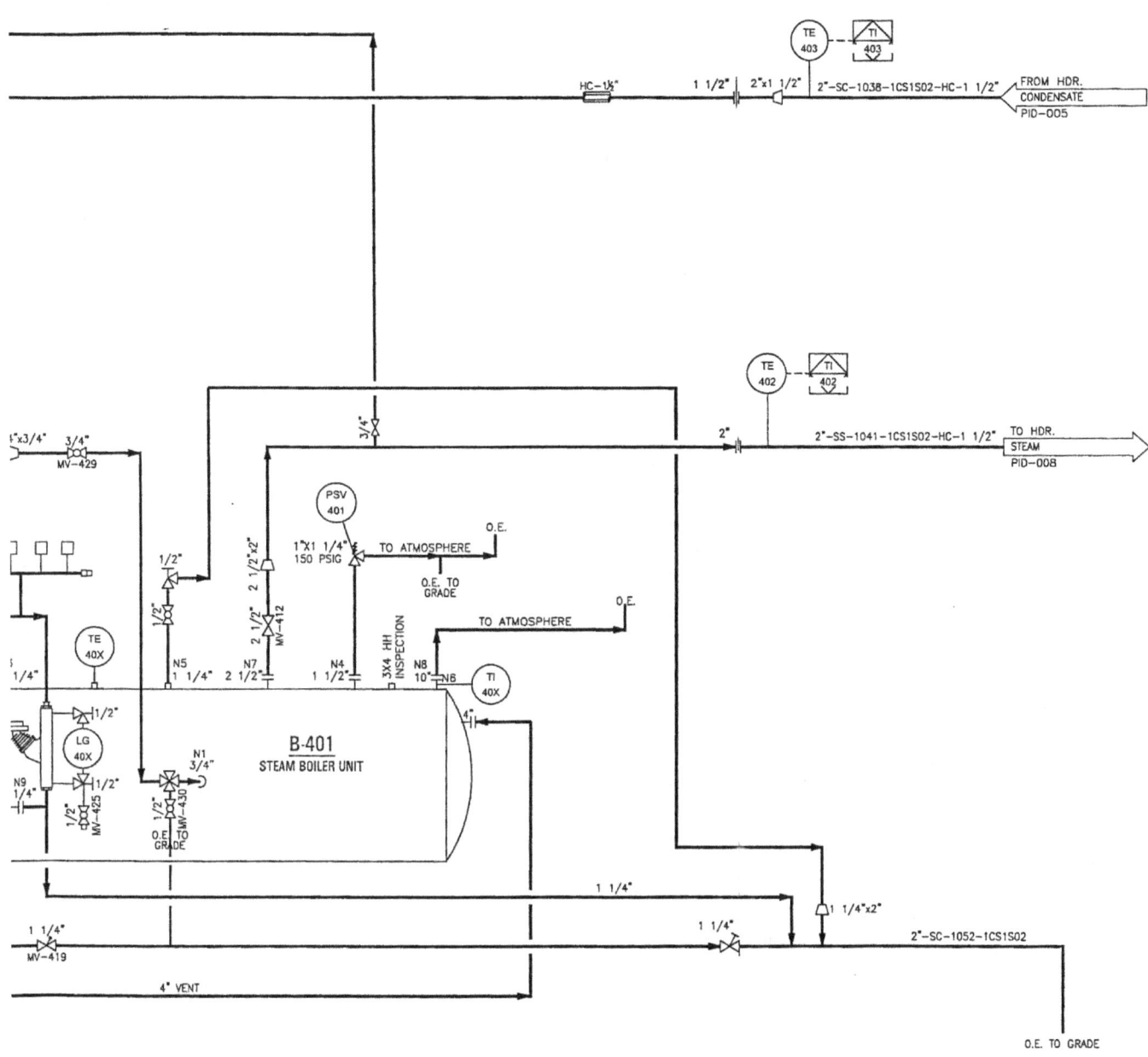

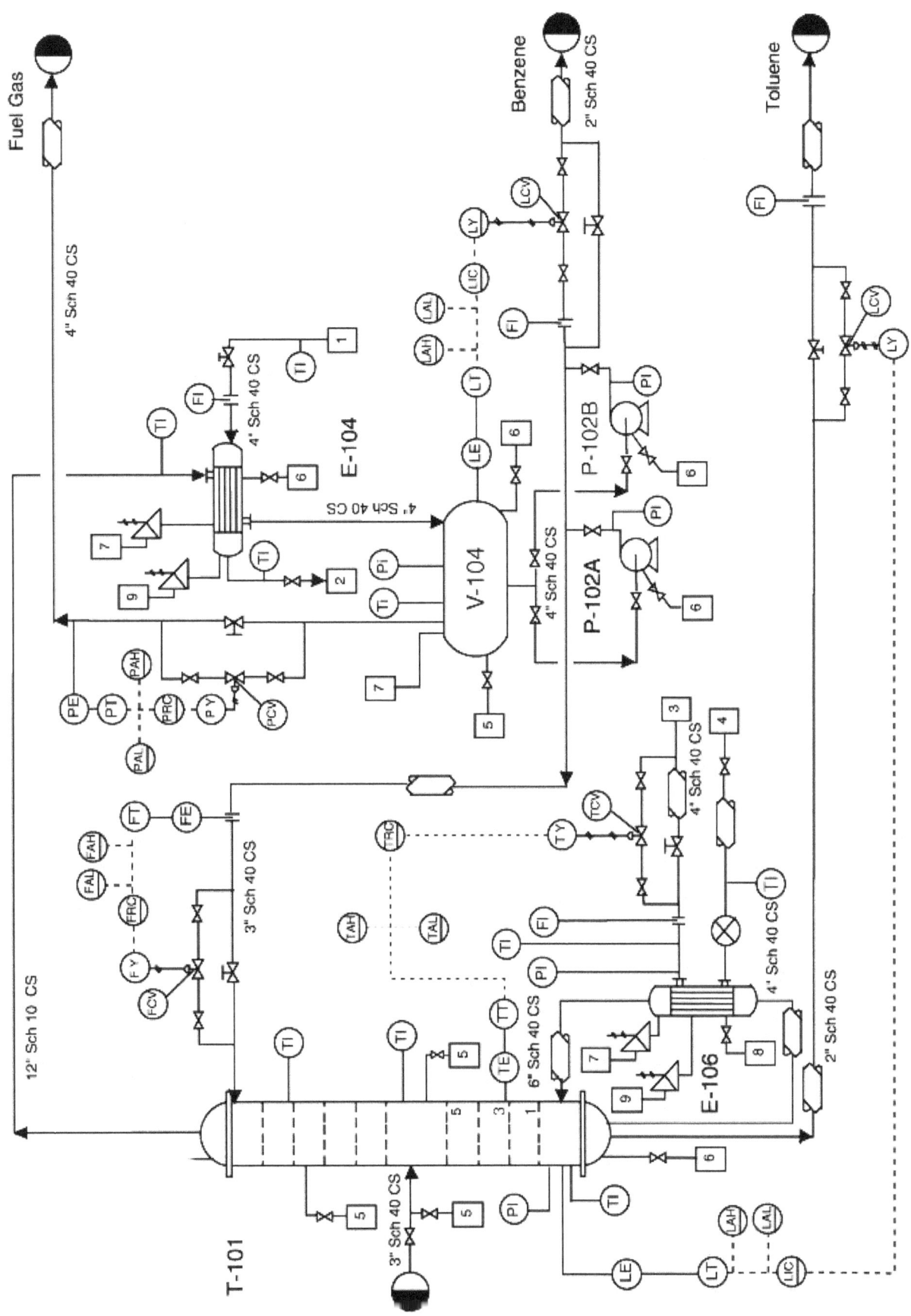

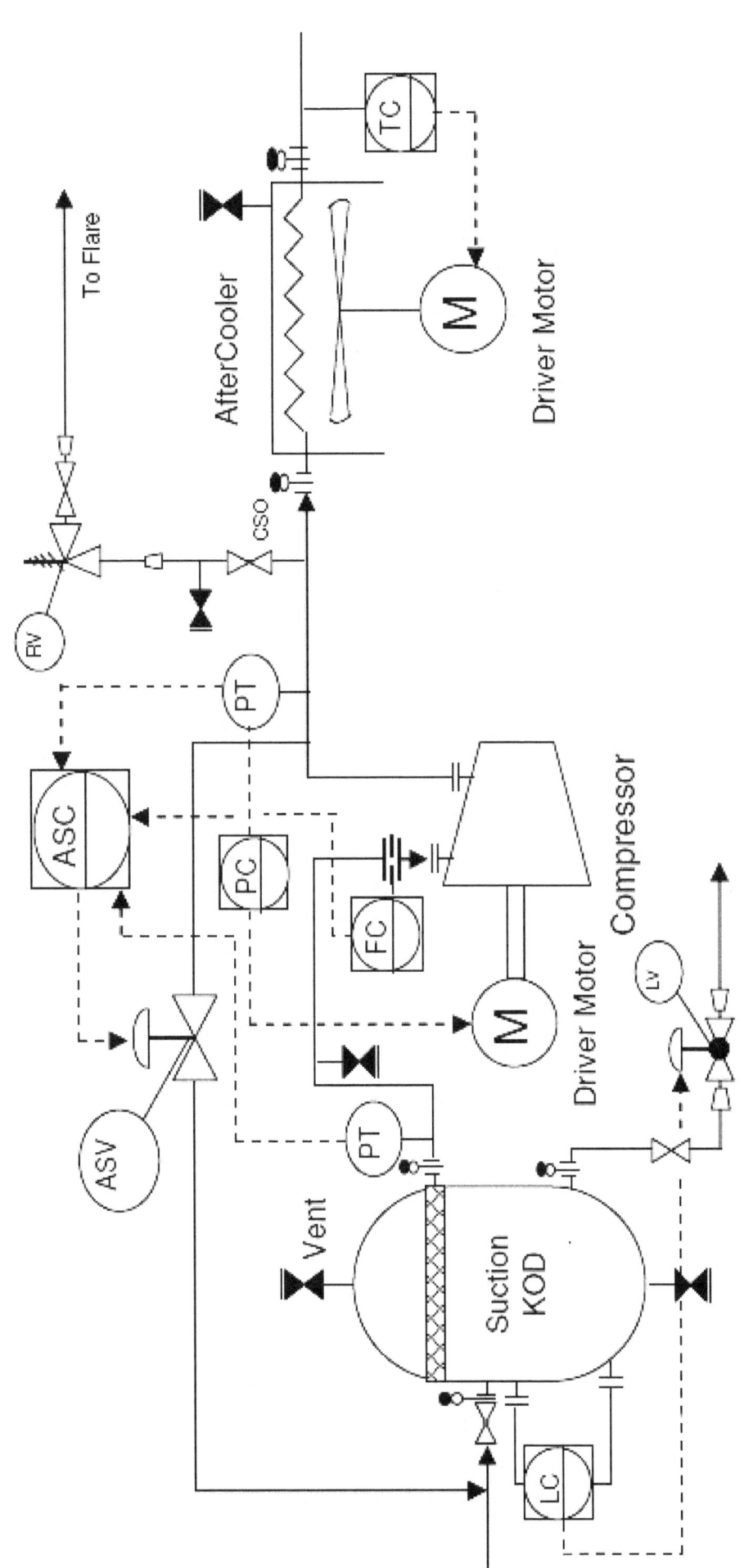

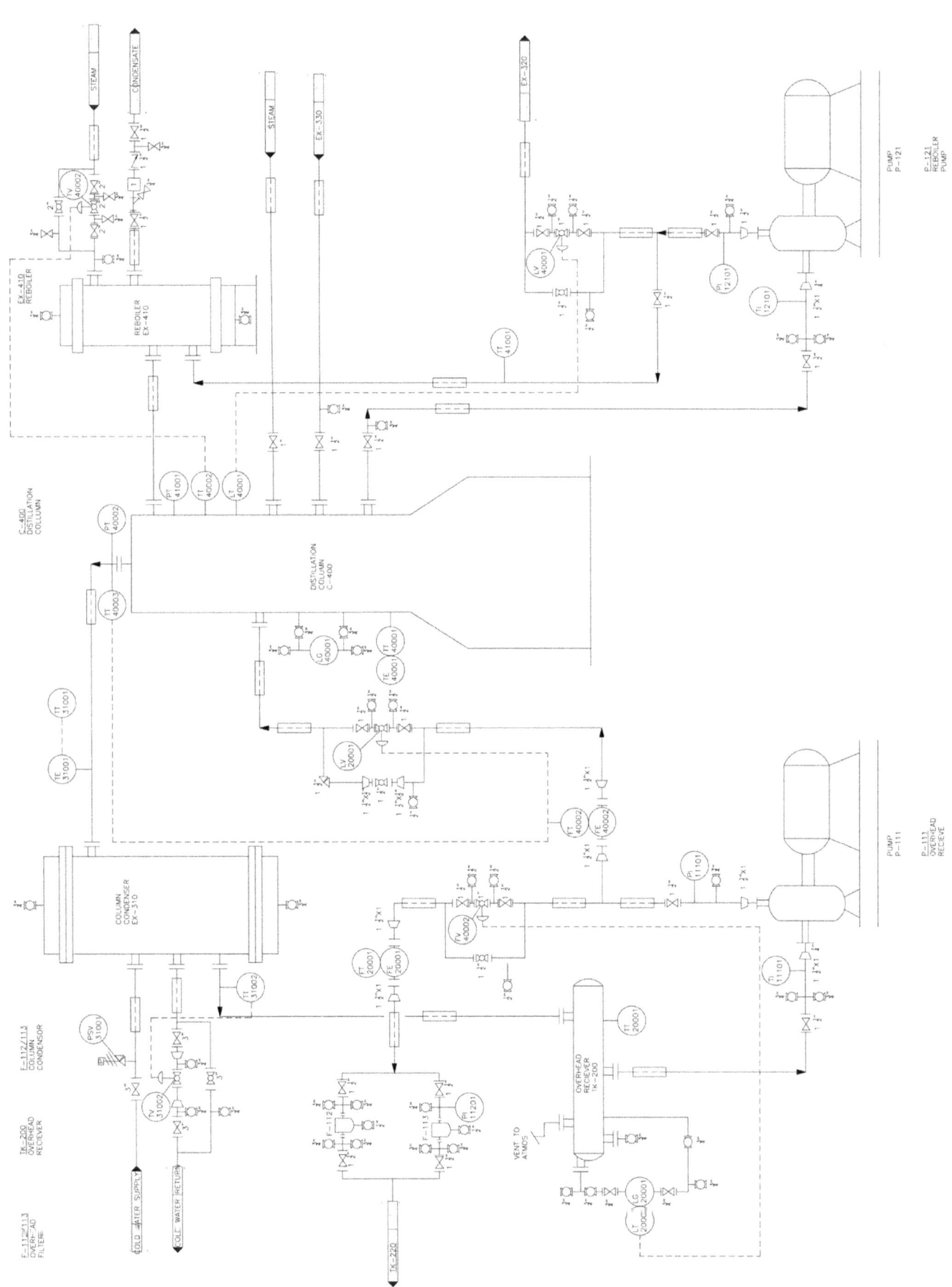

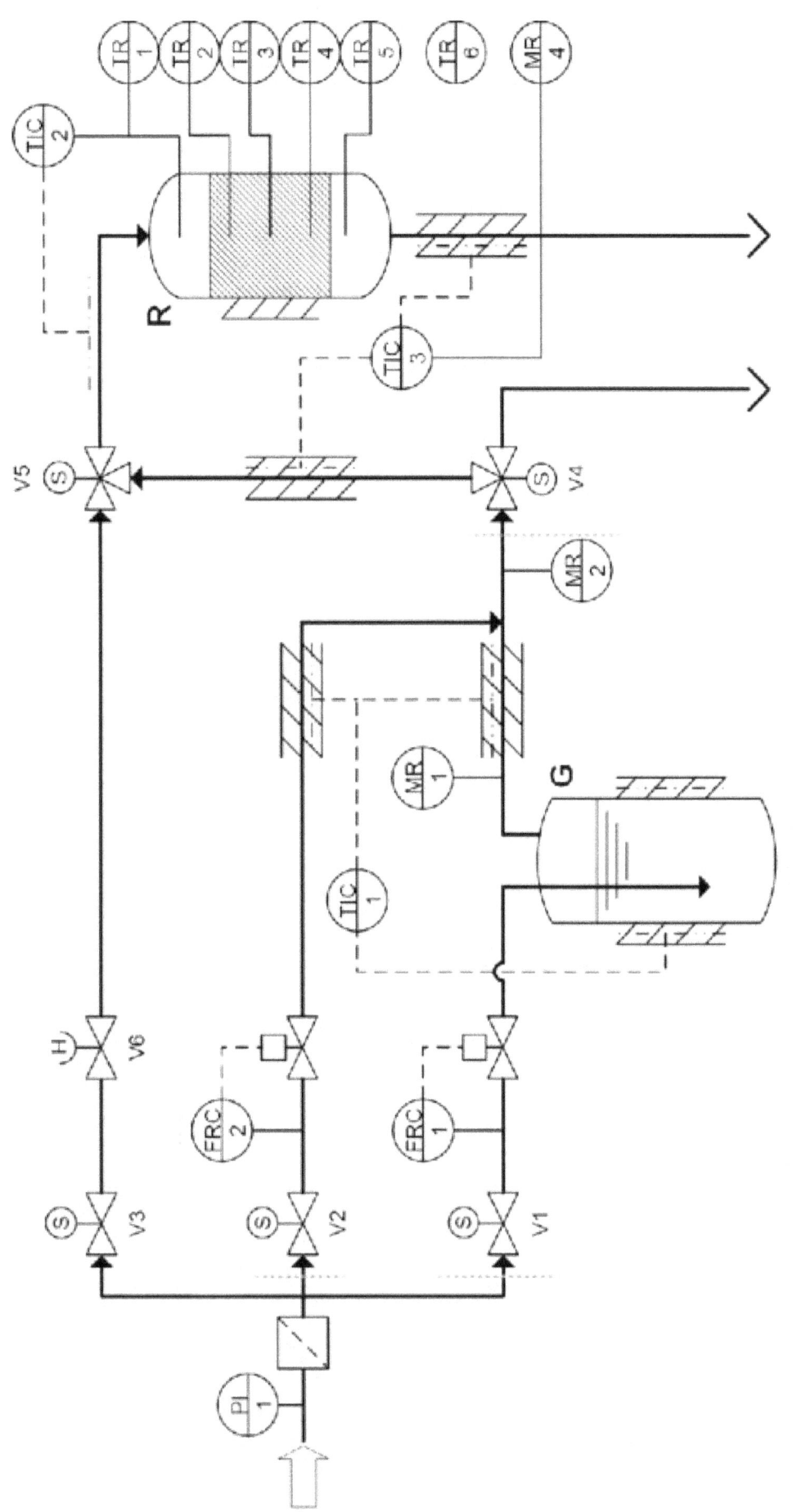

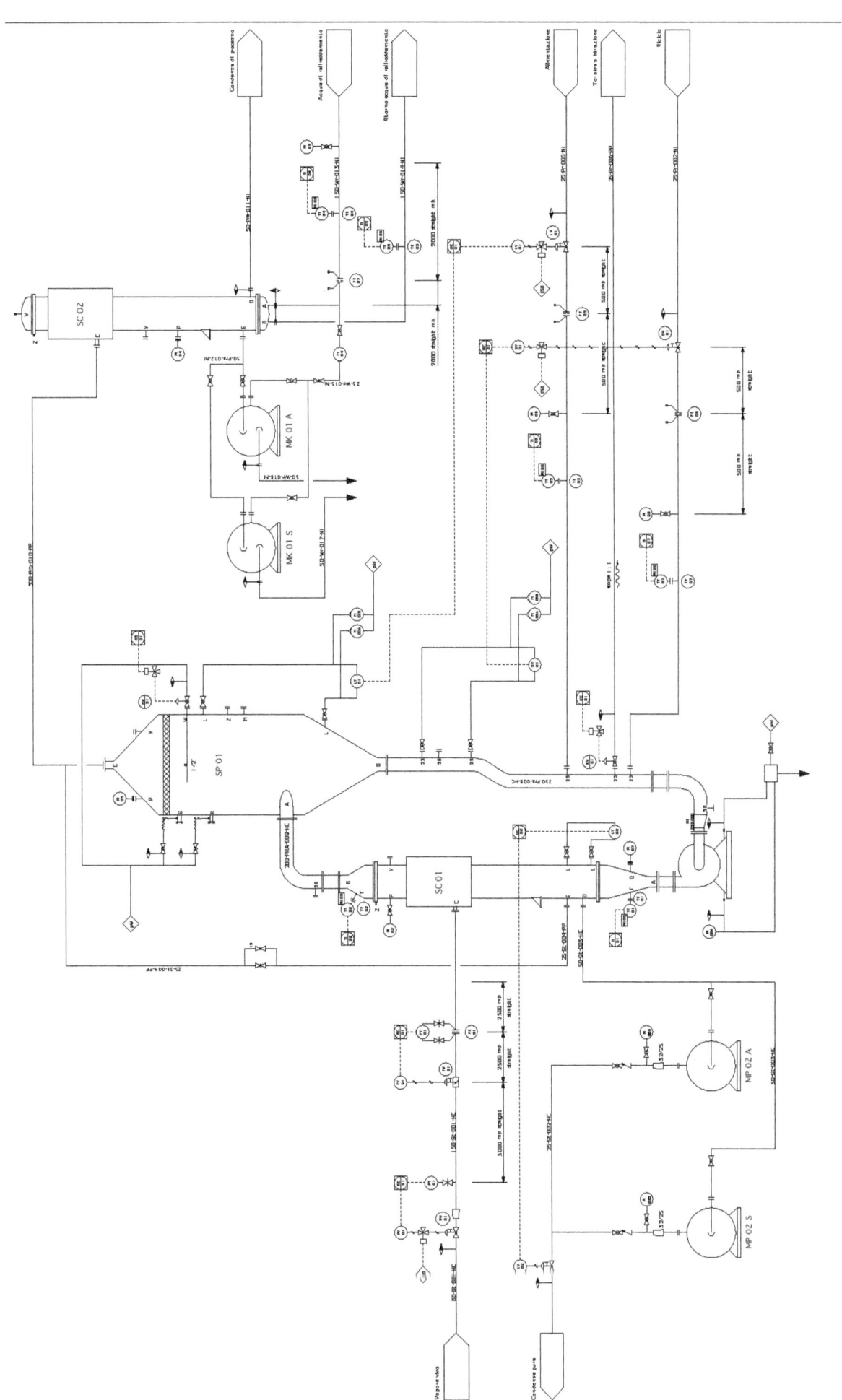

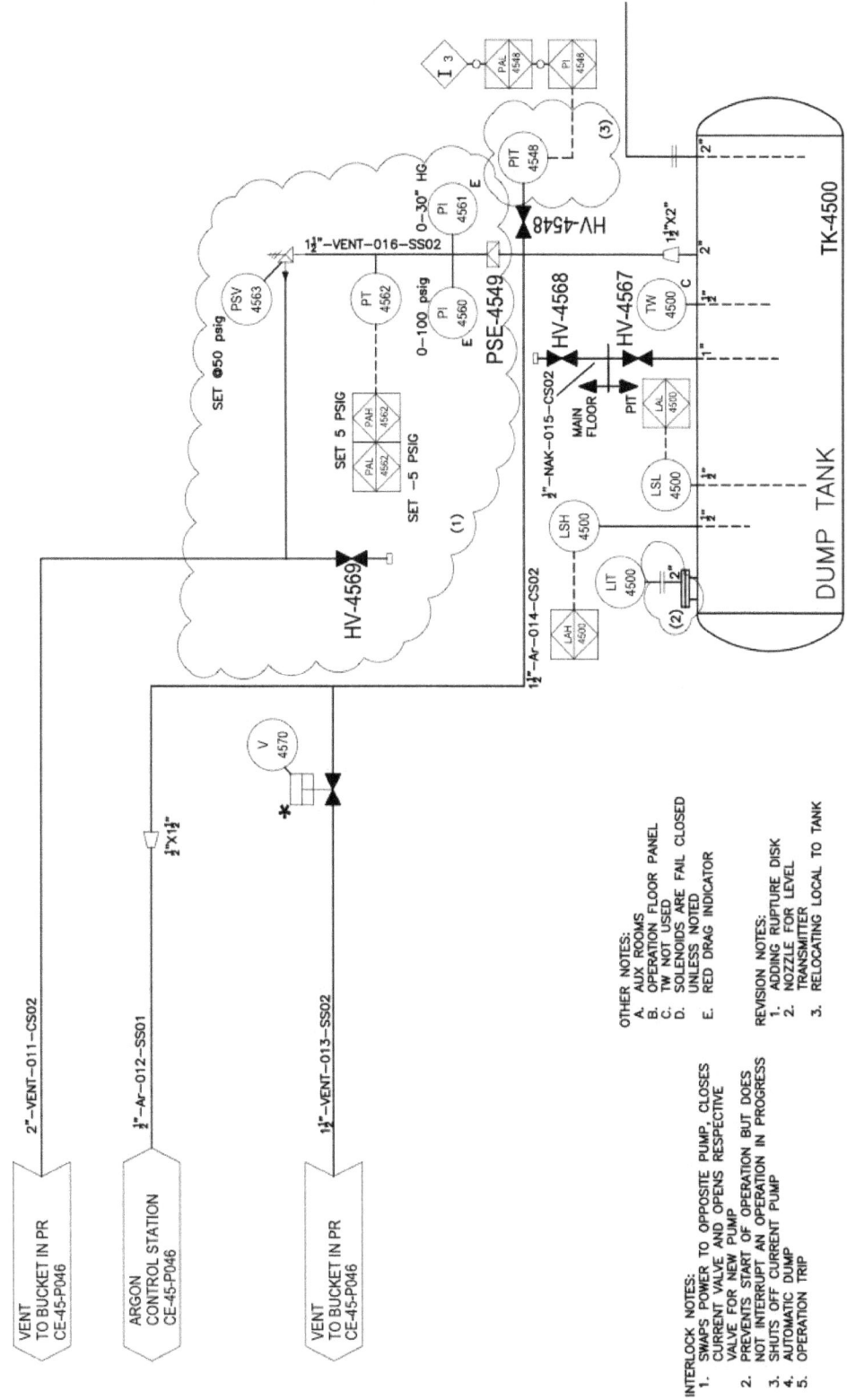

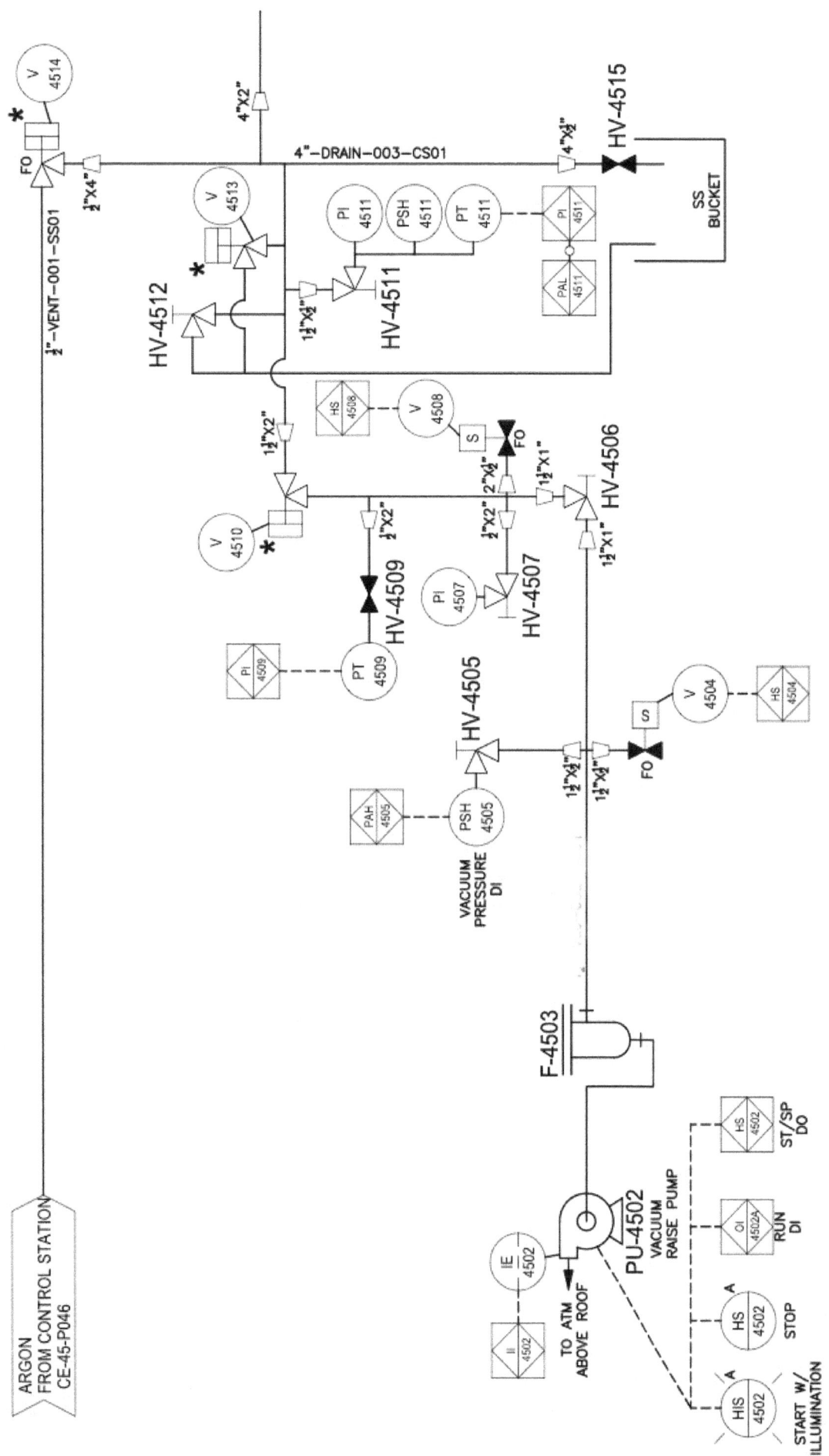

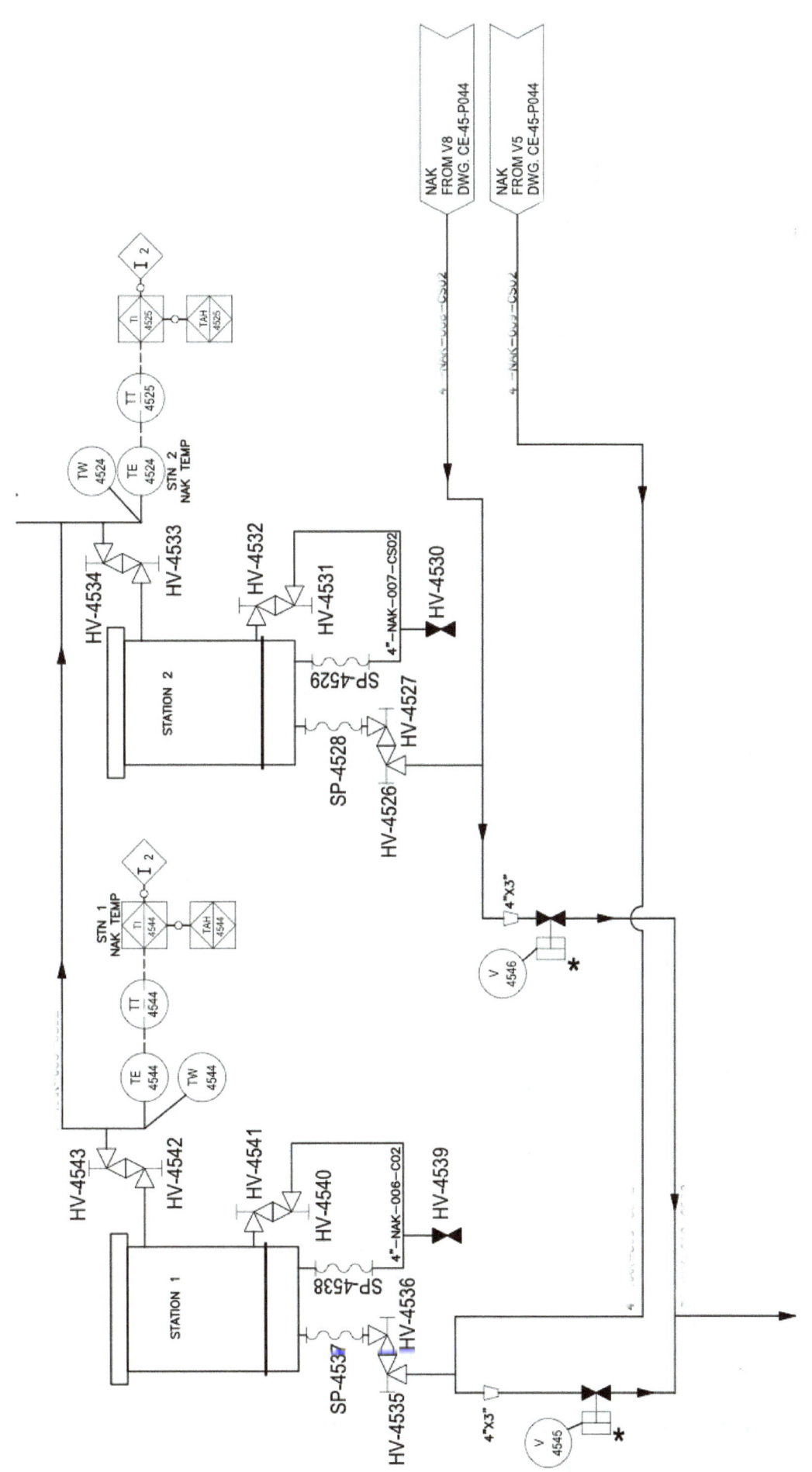

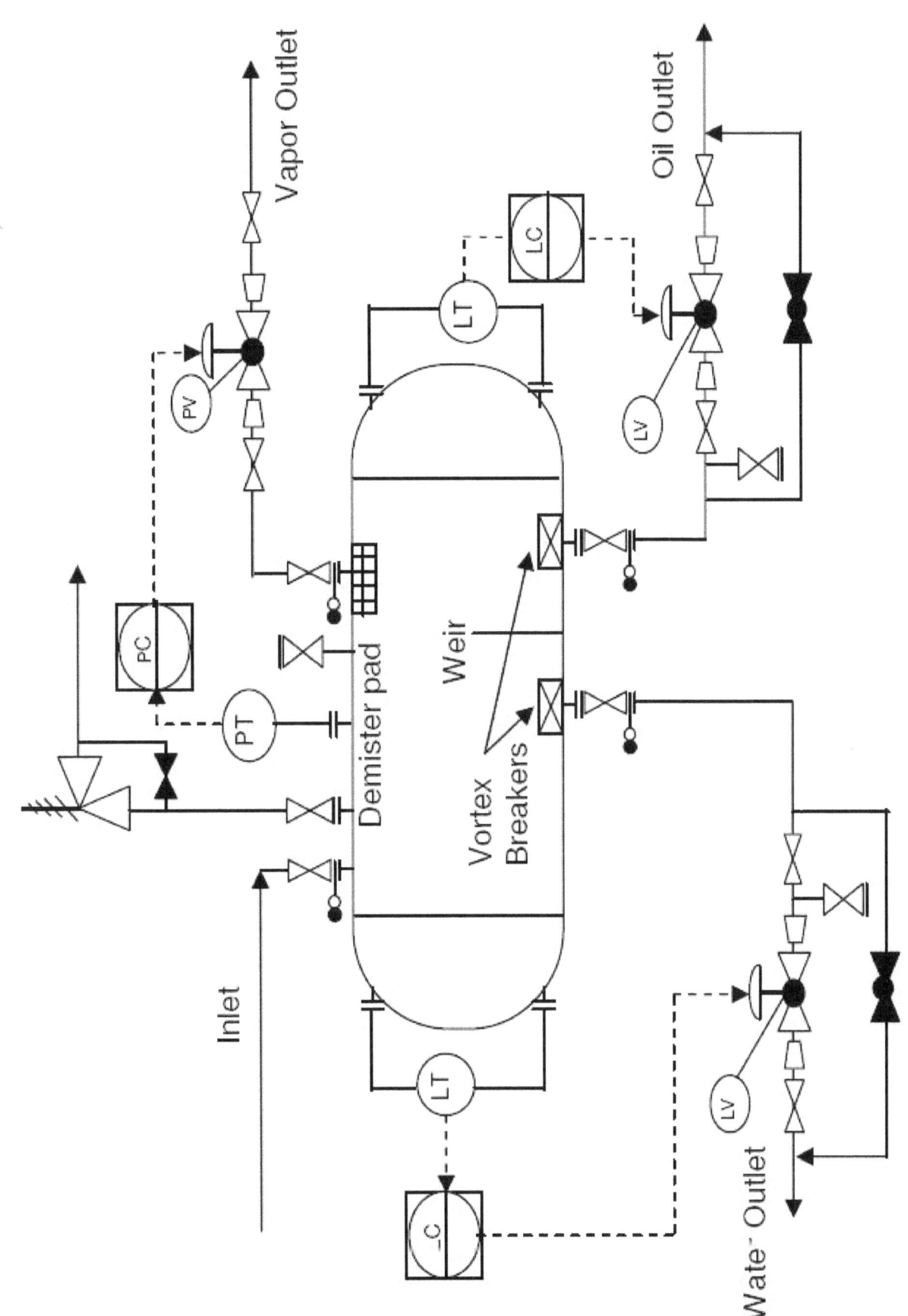

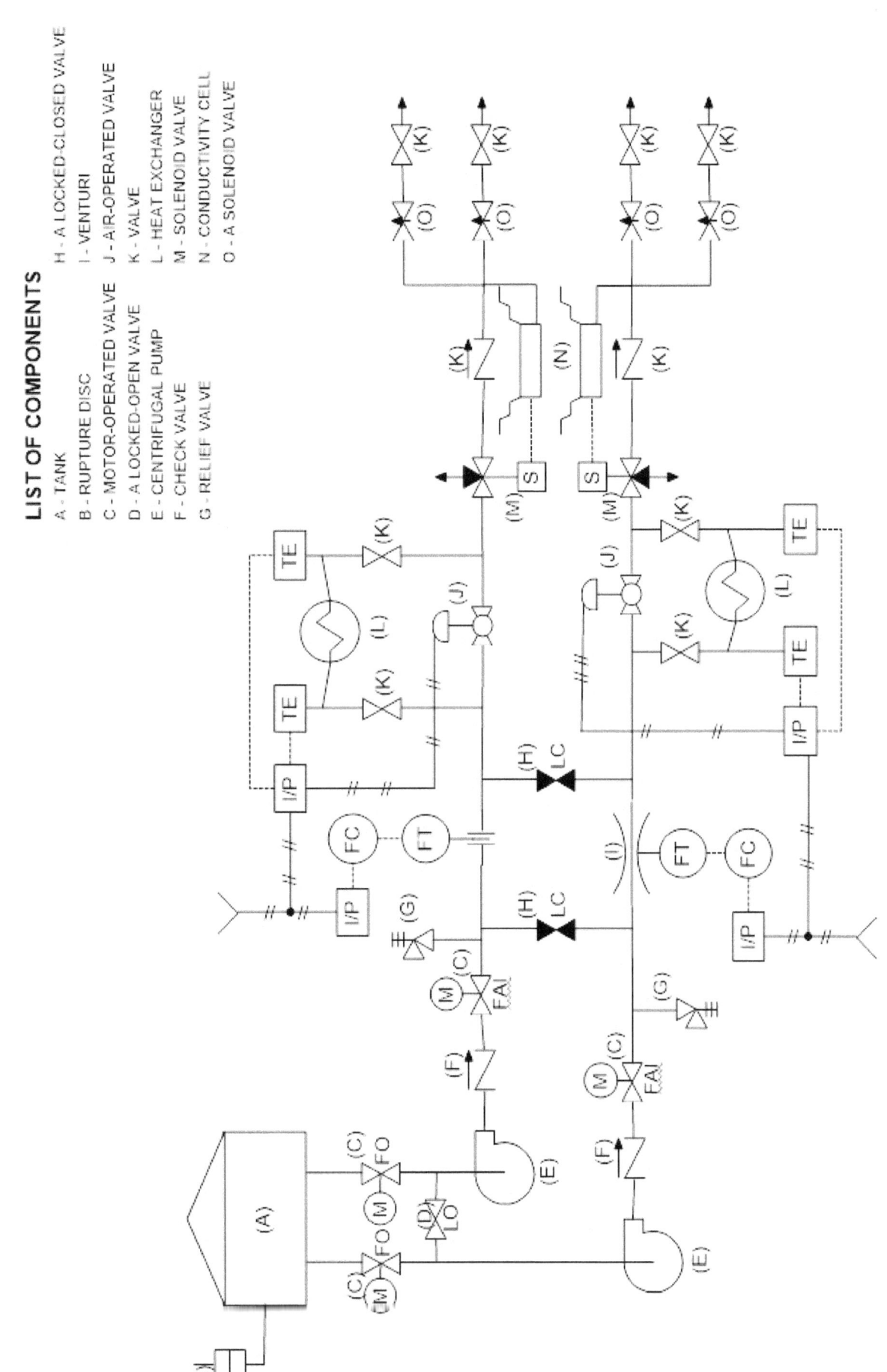

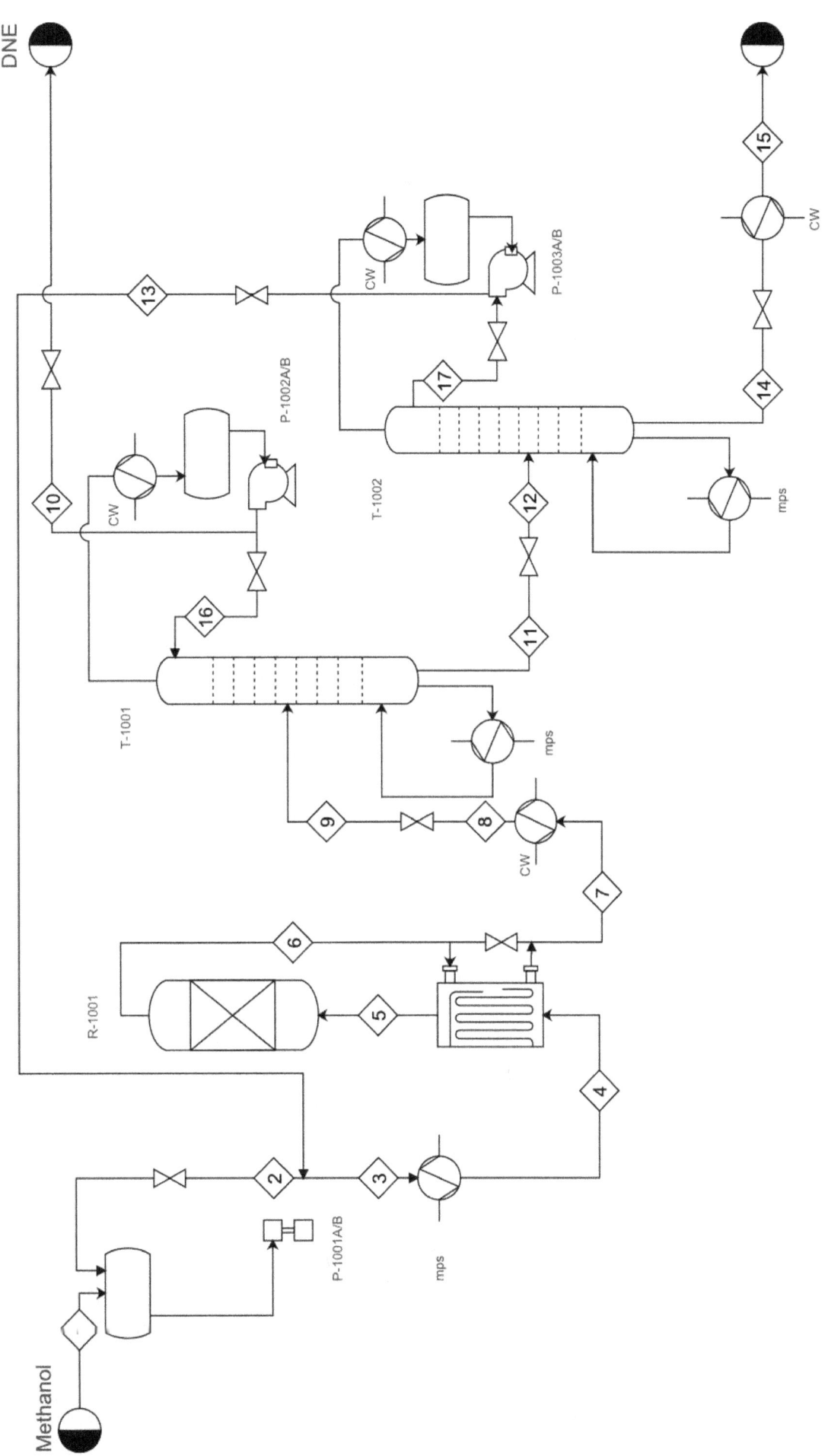

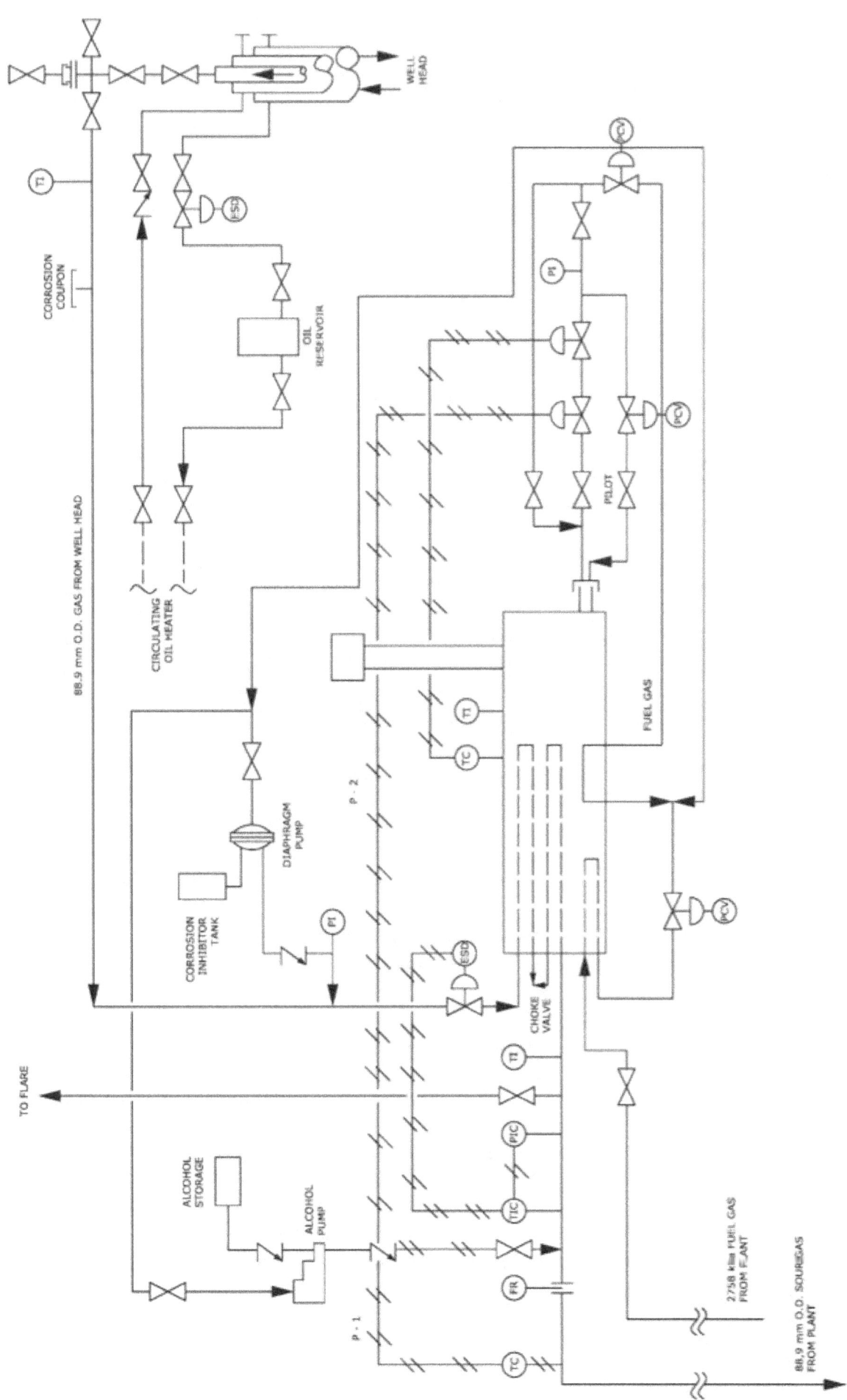

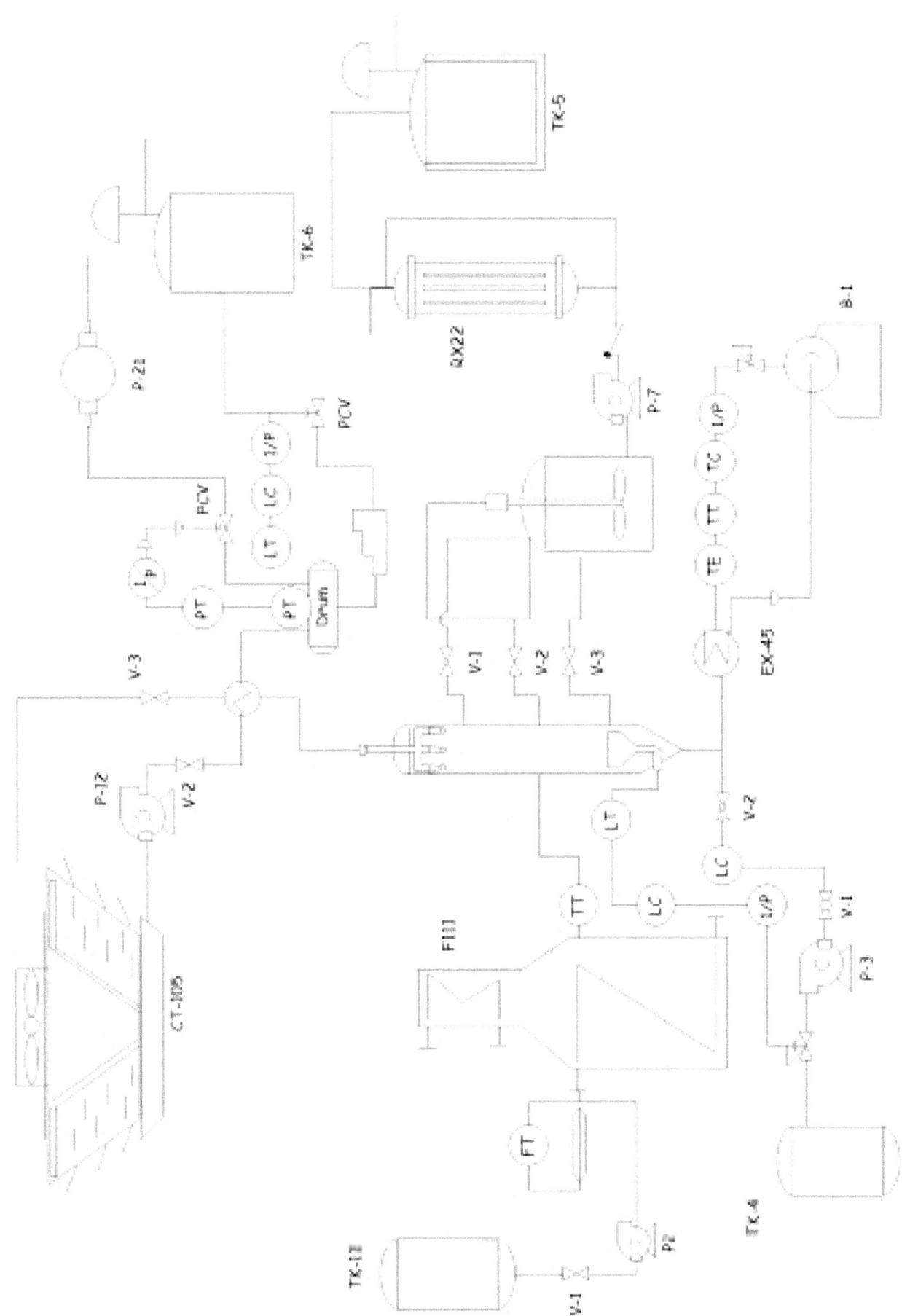

Delta-Wye

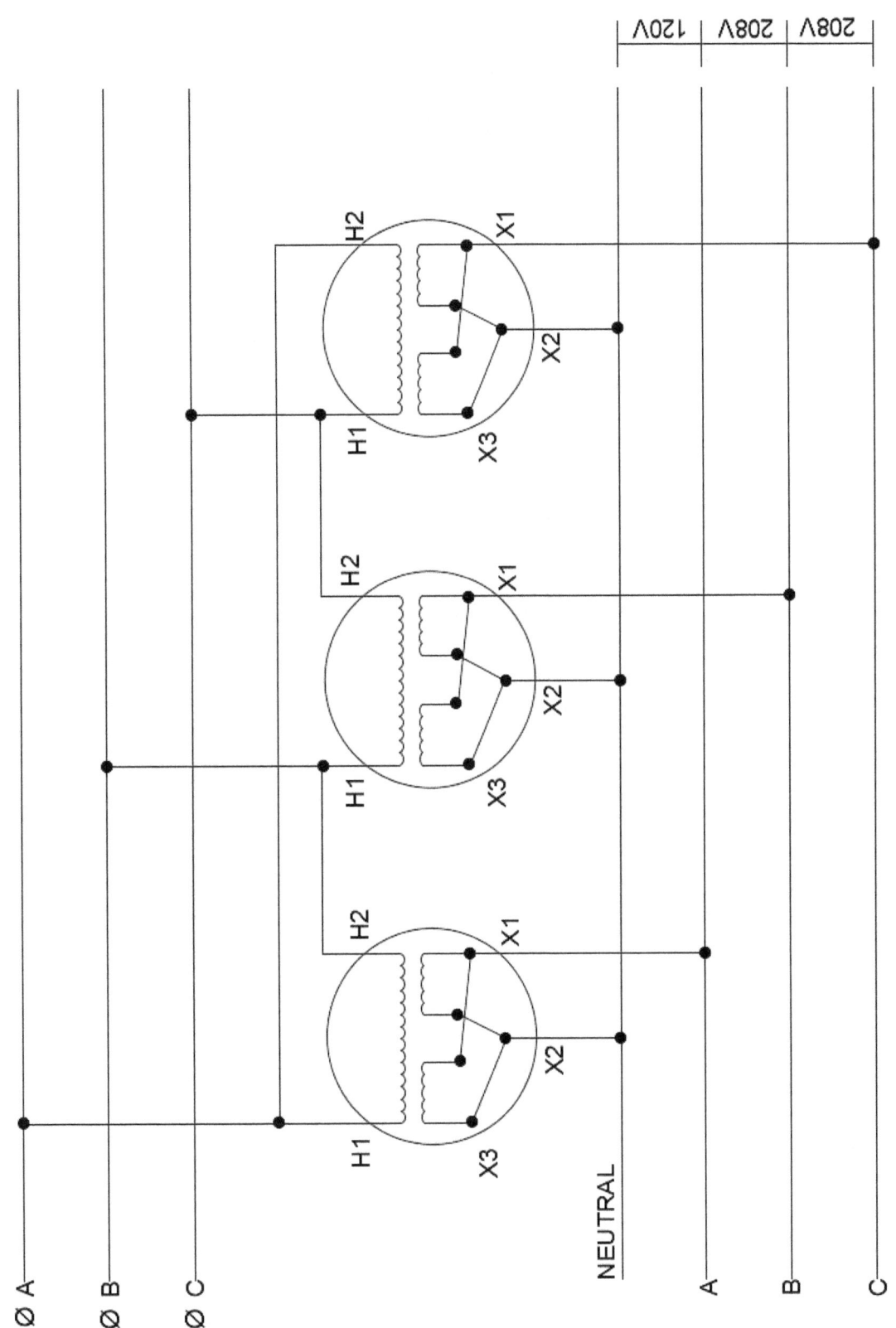

Wye-Wye

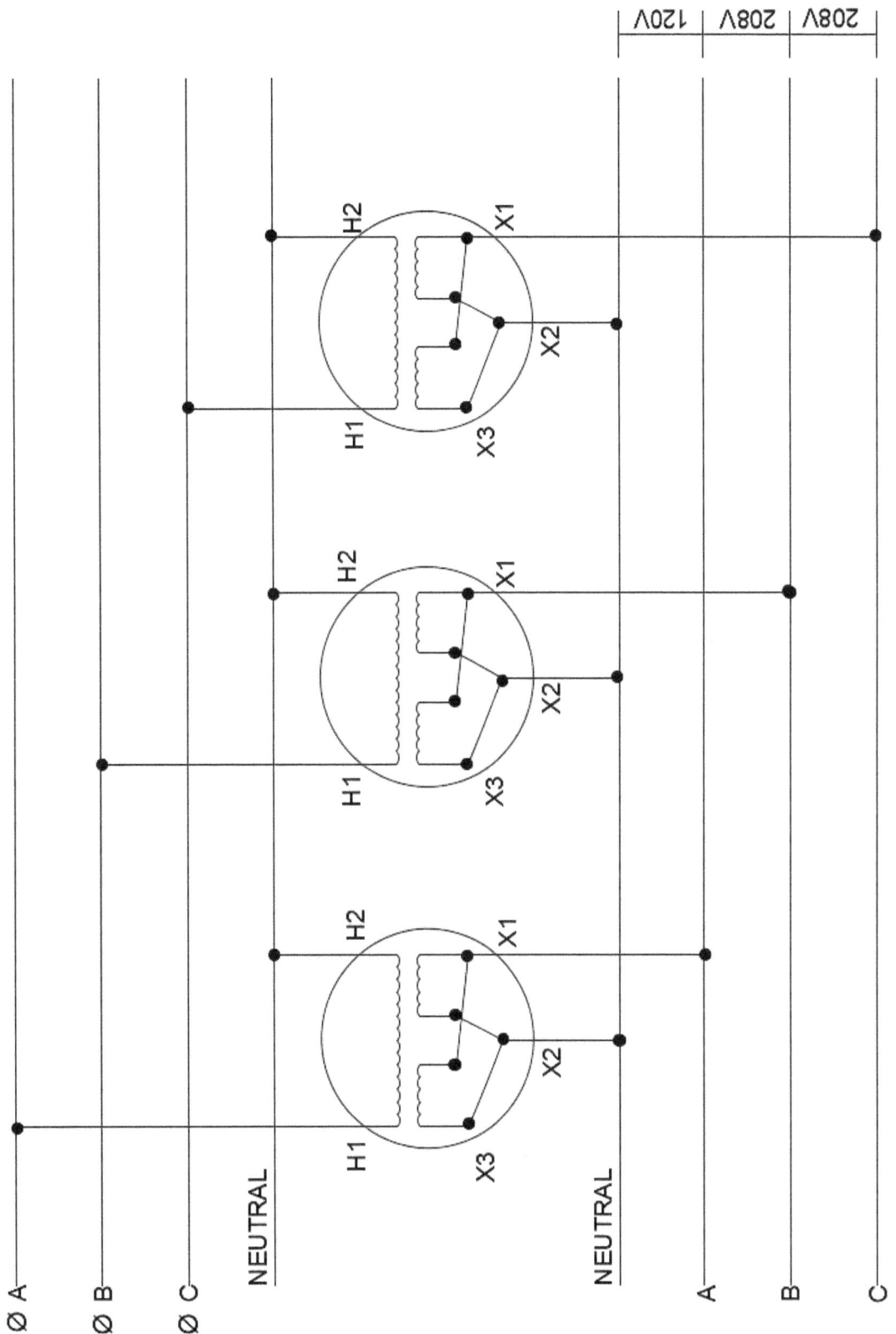

Delta-Delta

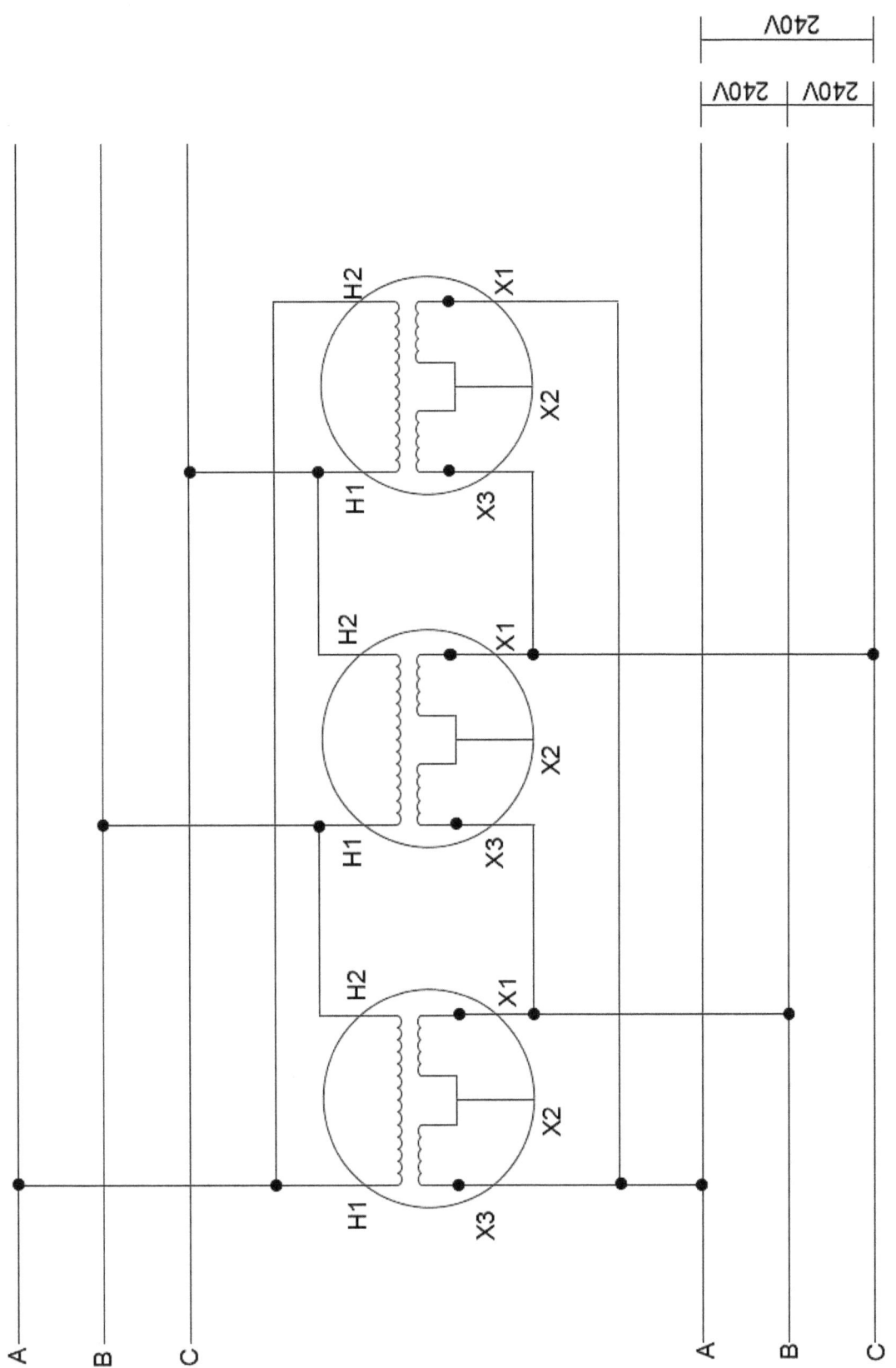

"Never judge your full potential

Based on your first run."

- *Chip Gaines*

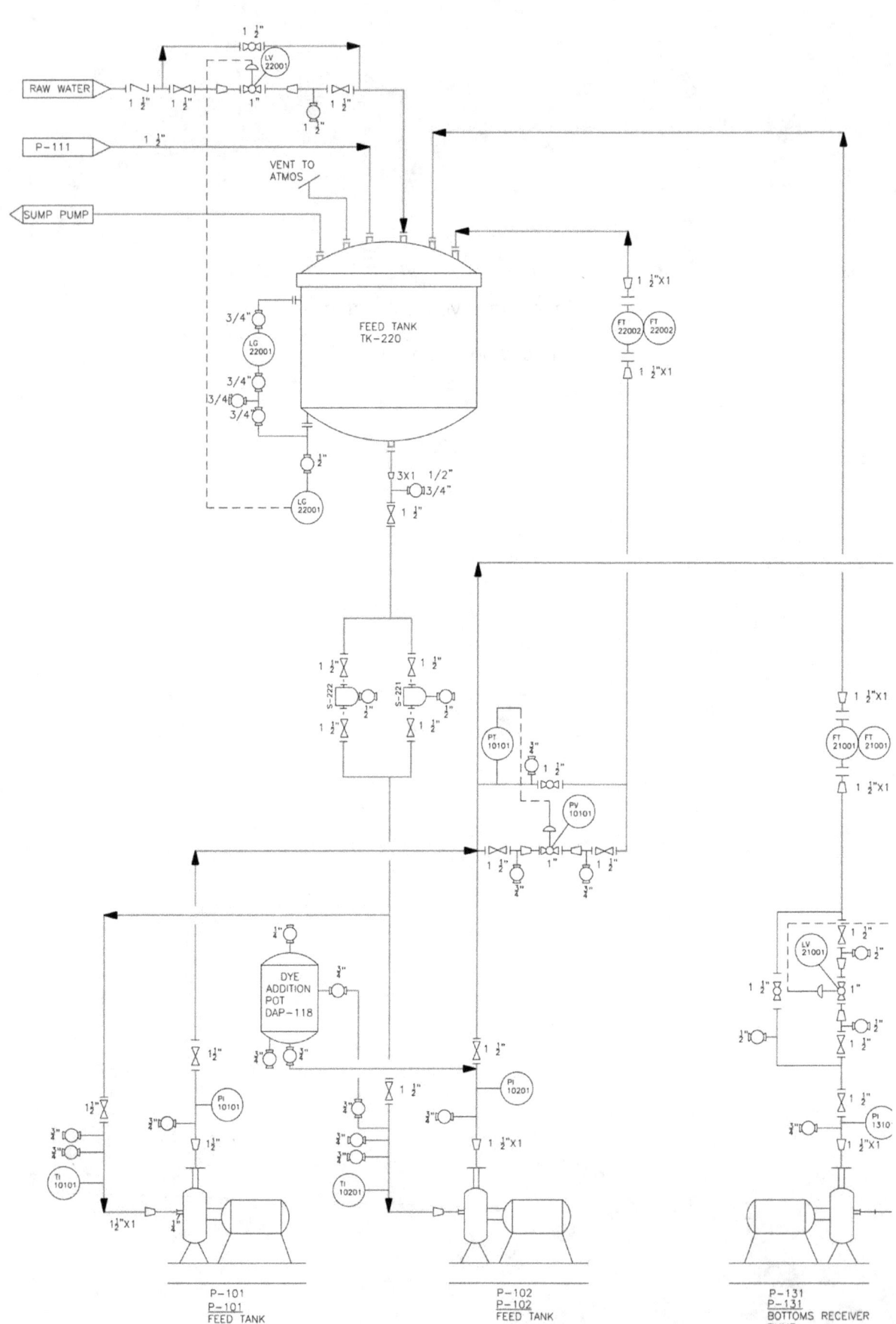

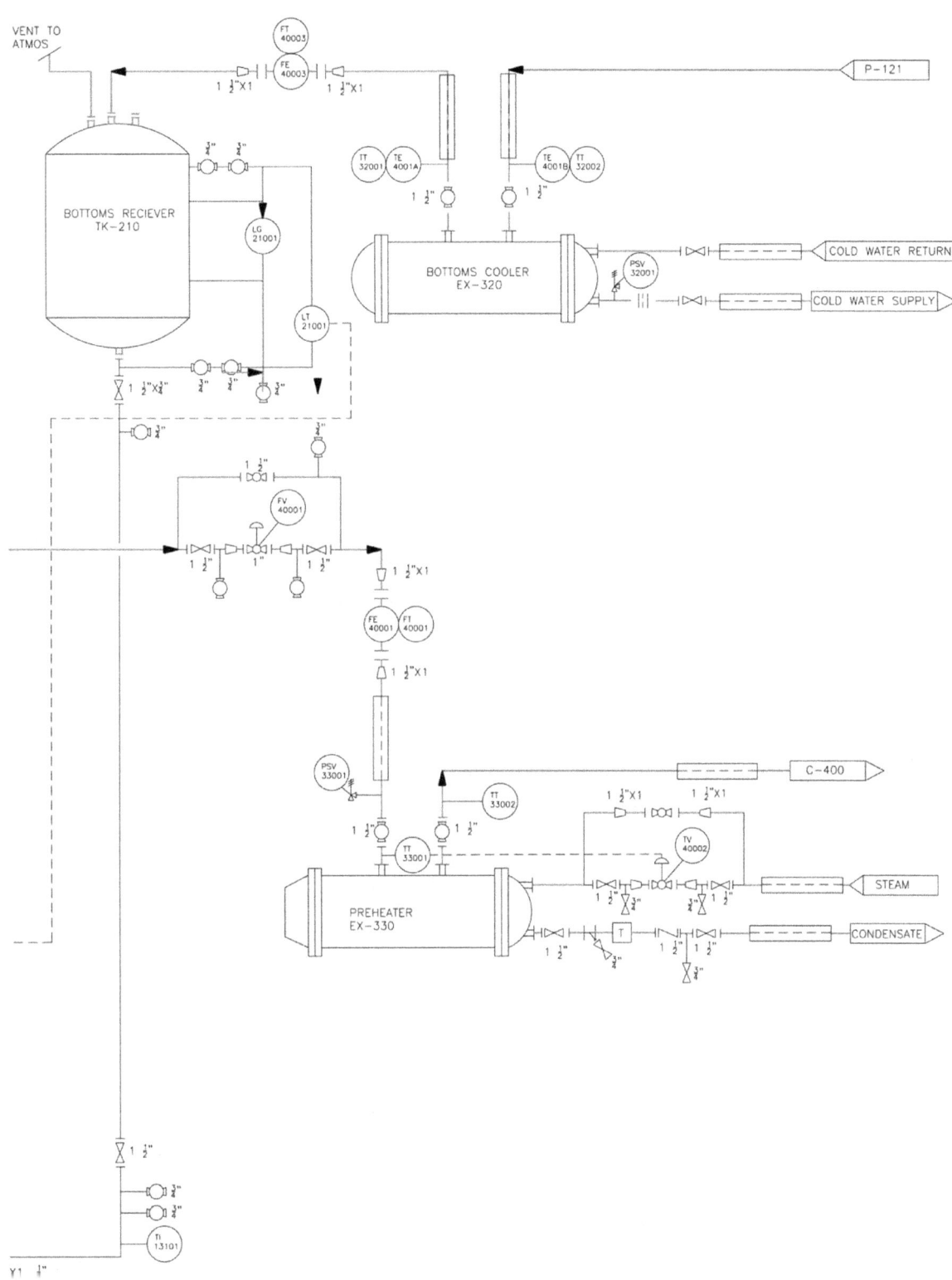

"If you trust in yourself...
And believe in your dreams...
And follow your star...
...you'll still get beaten by people who spent their time working hard and learning things and weren't so lazy."

- *Terry Pratchett*

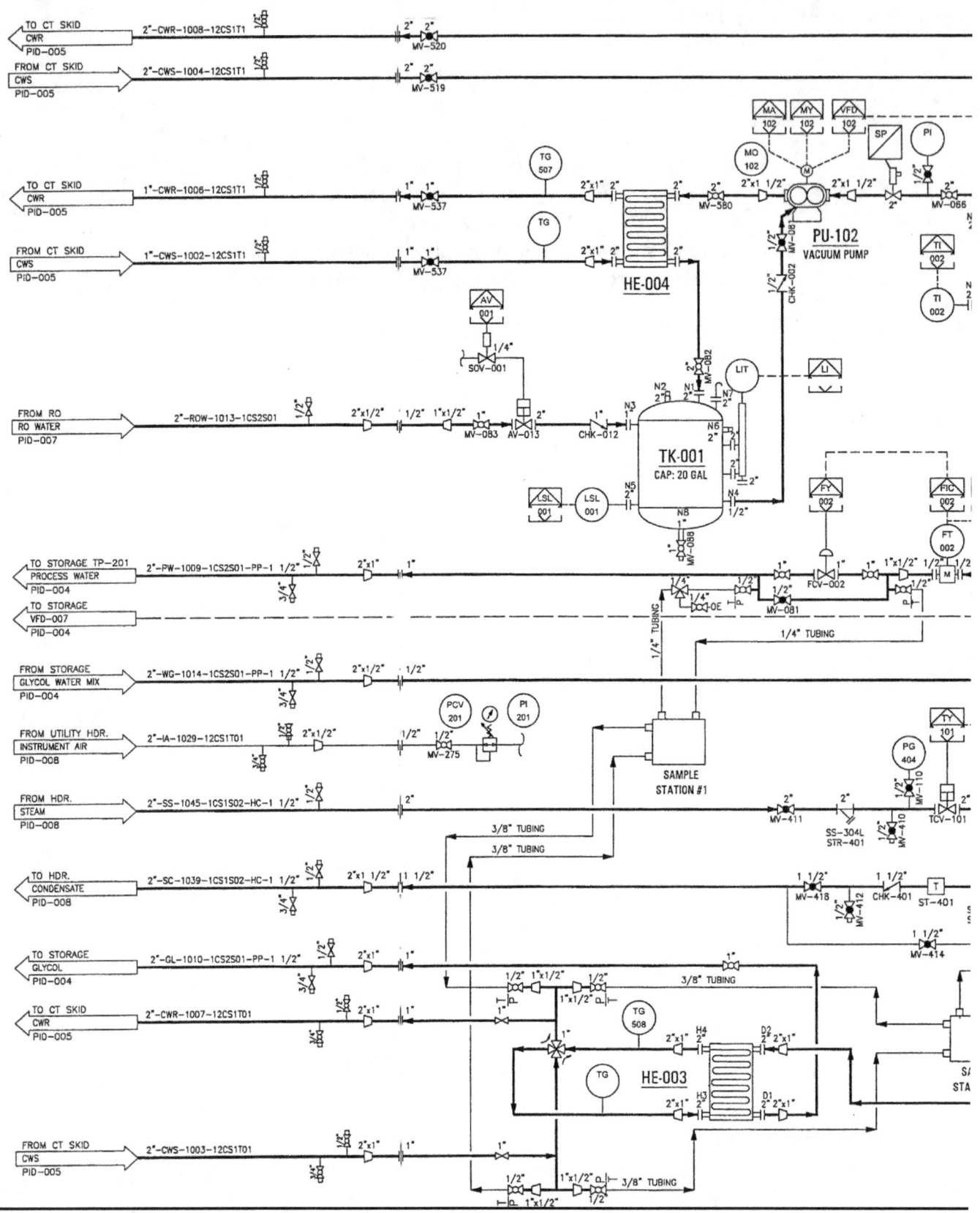

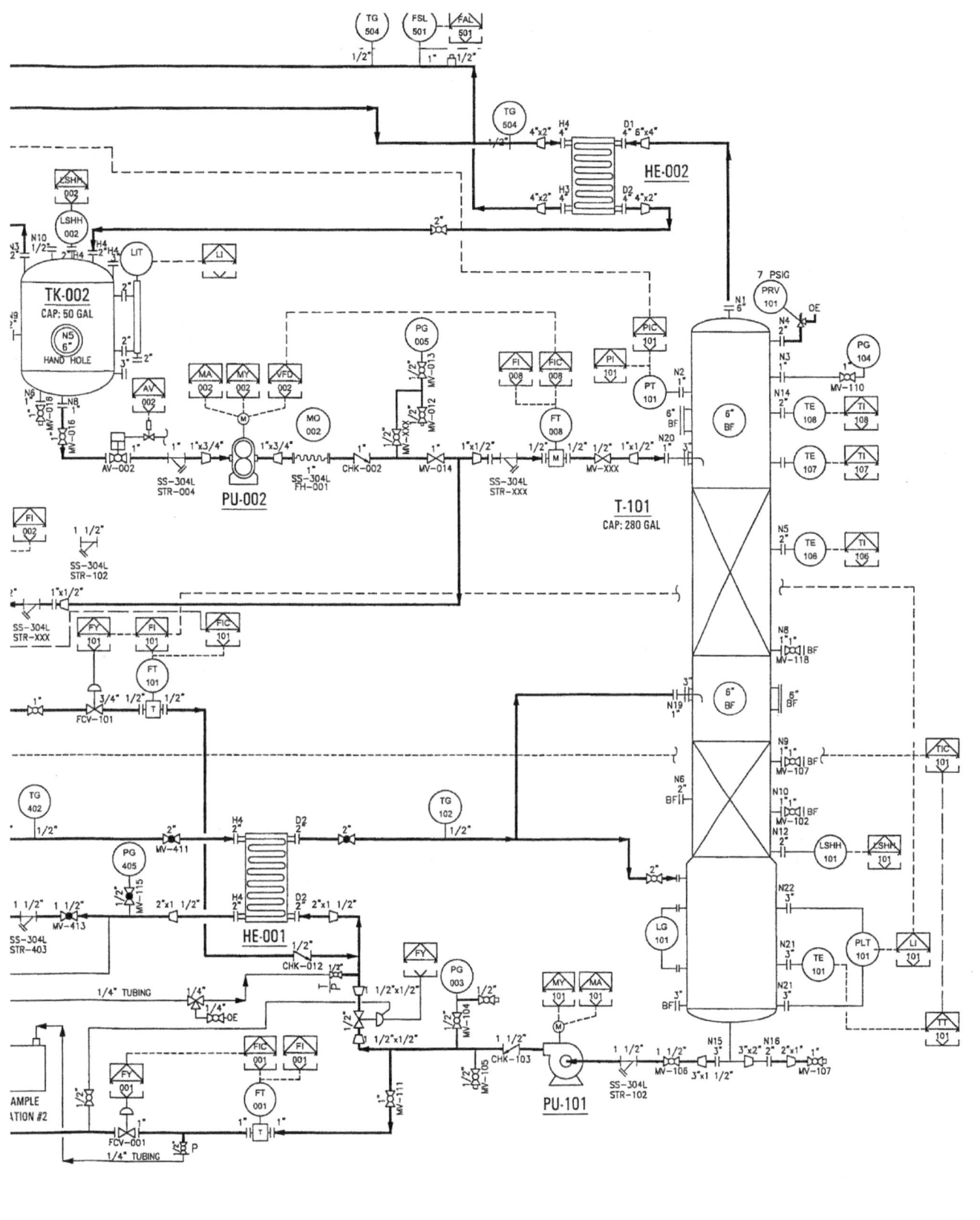

"It is possible to commit **no** mistakes and still lose. That is **not** a weakness; that is life."

- *Jean-Luc Picard*

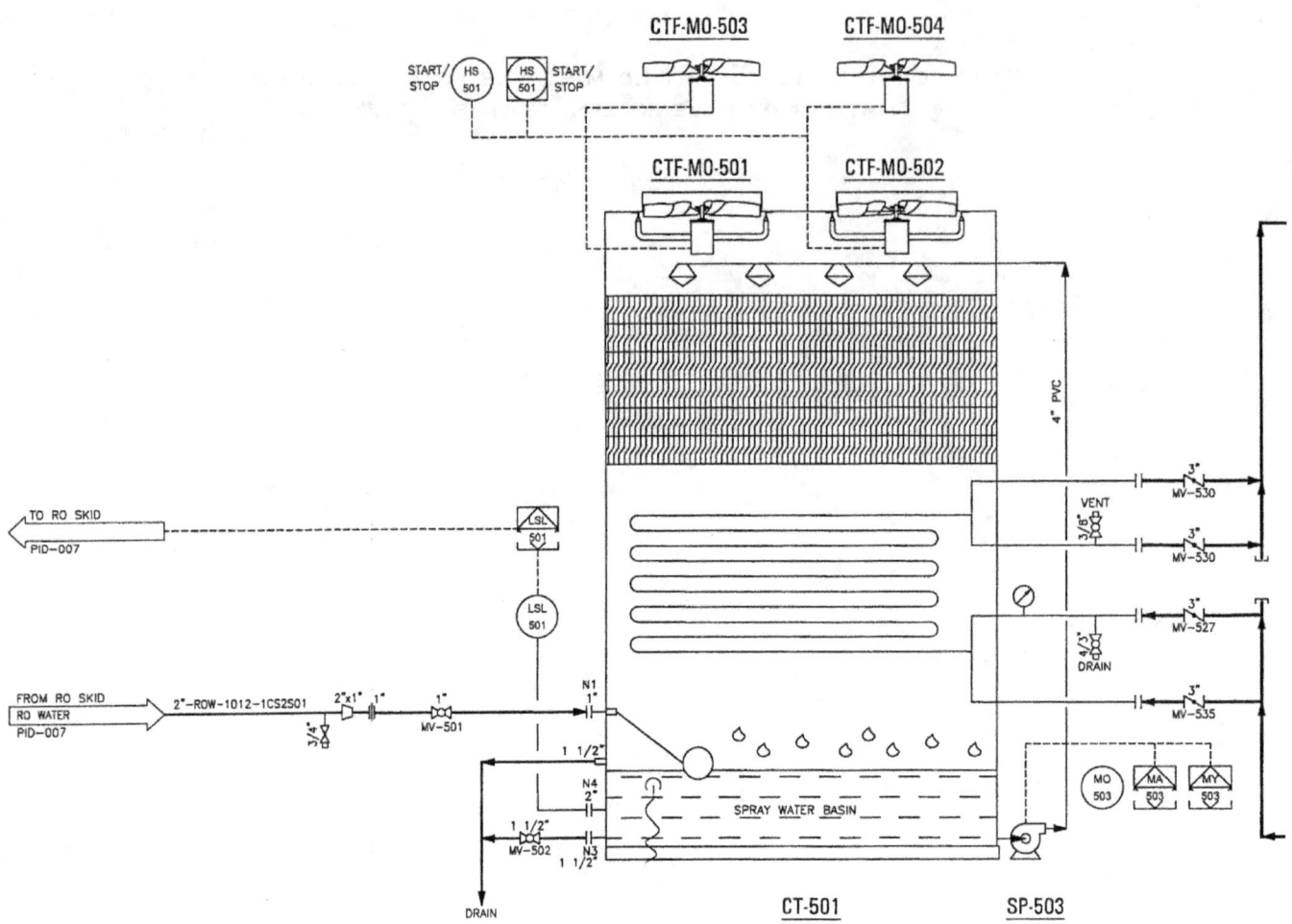

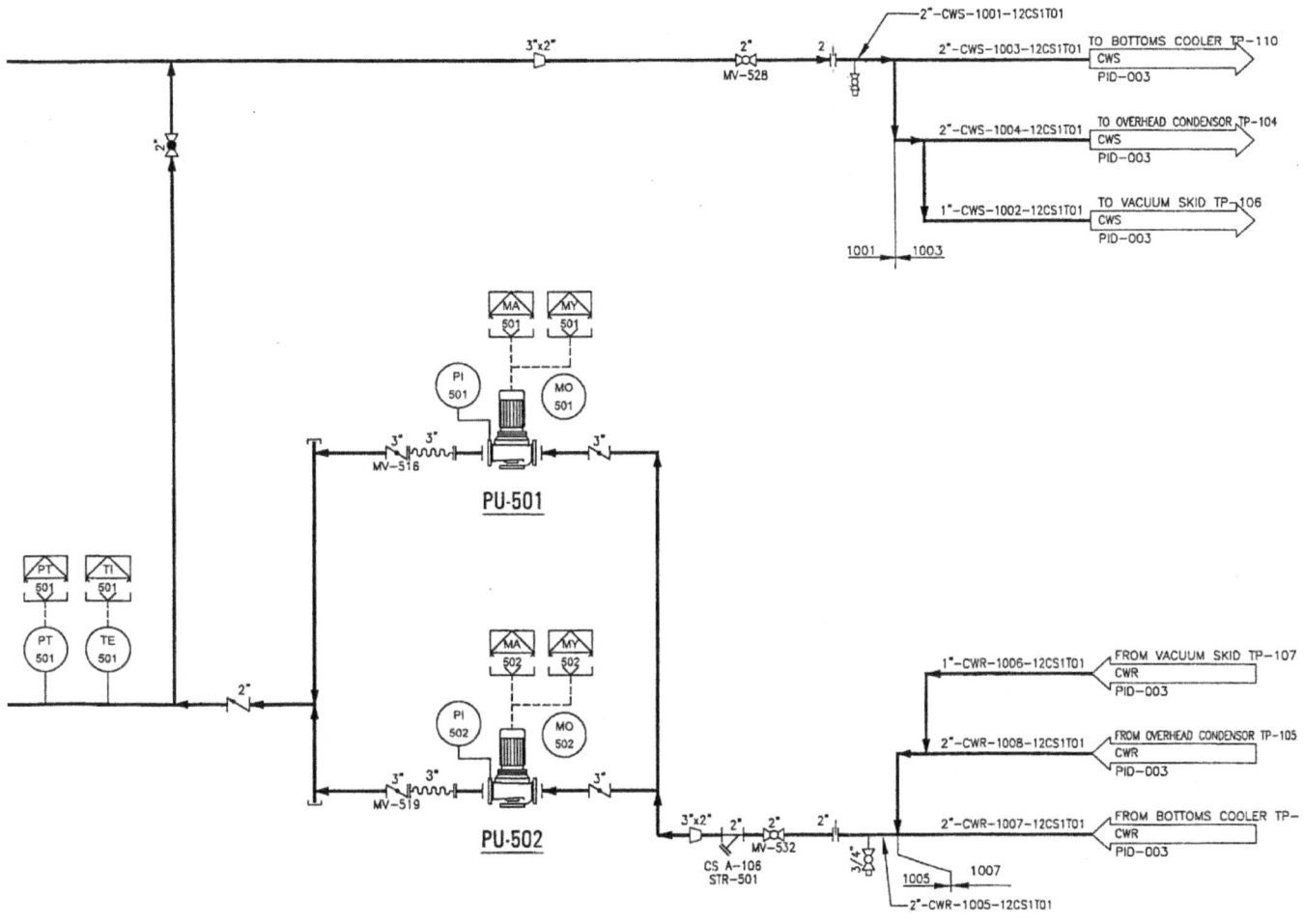

"Tomorrow is the most important thing in life.
Comes into us at midnight very clean.
It's perfect when it arrives and it puts itself in our hands.
It hopes we've learned something from yesterday."

- *Marion Morrison*

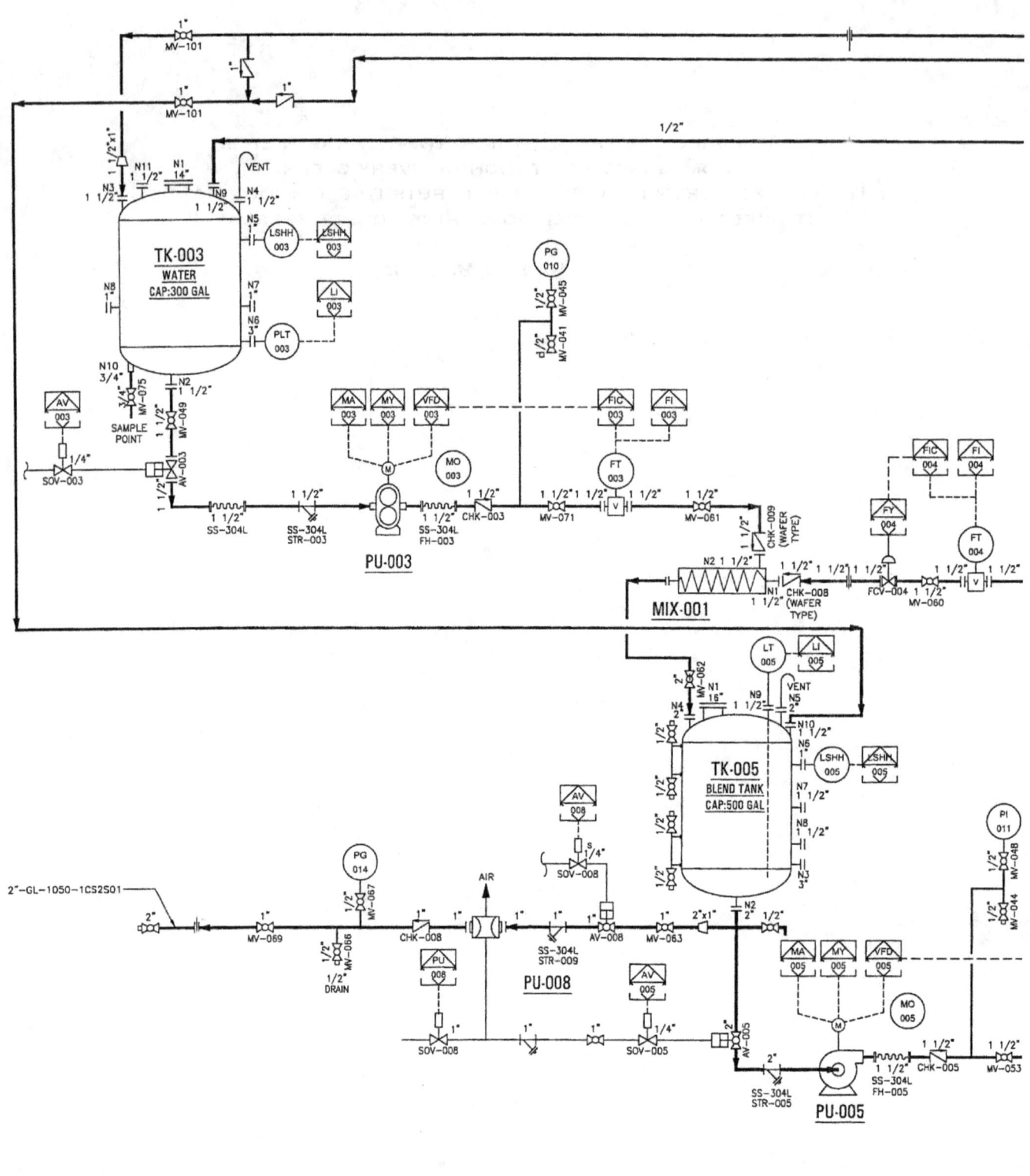

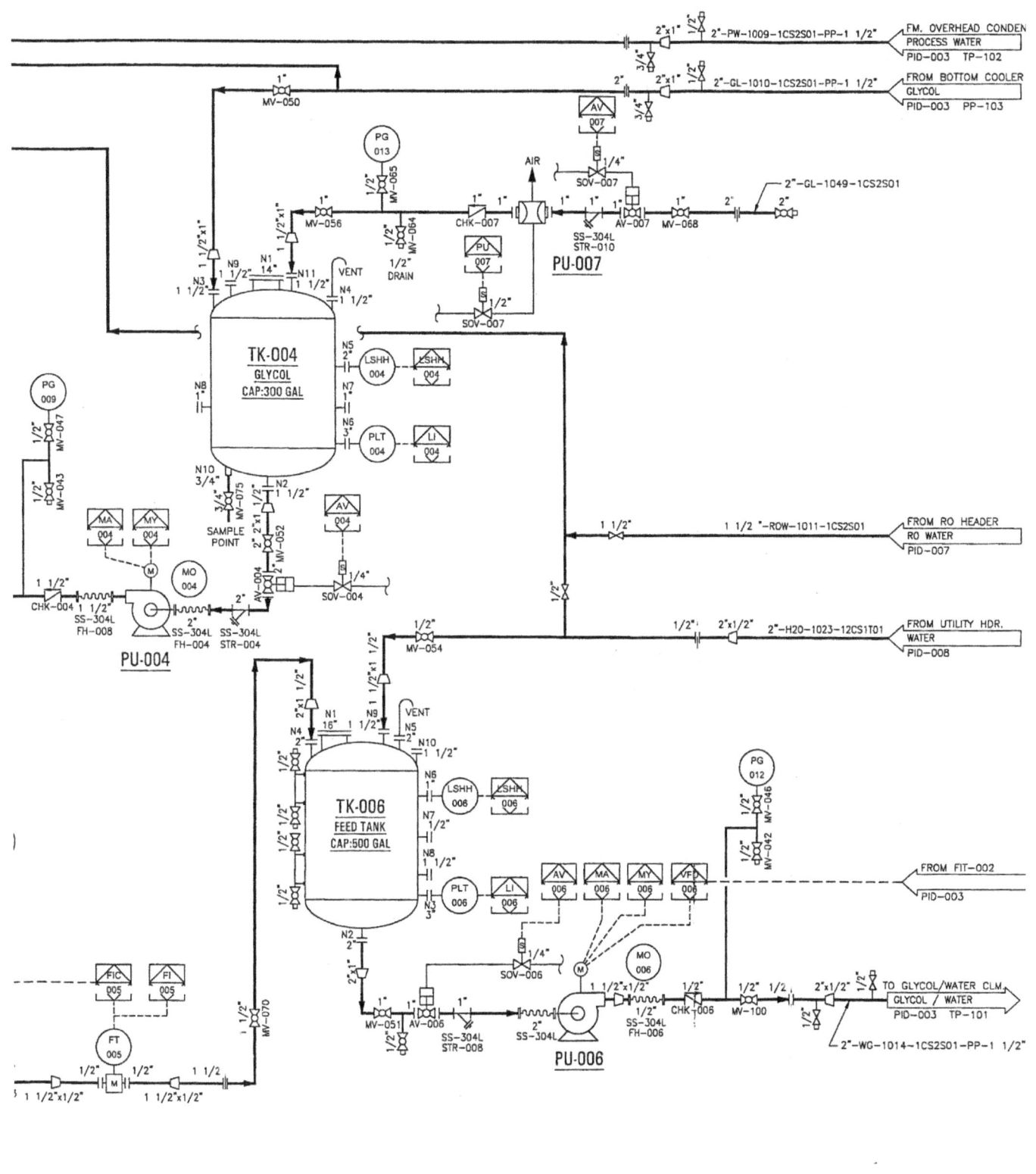

"I don't like looking back.
I'm always constantly looking forward.
I'm not the one to sort of sit and cry over spilt milk.
I'm too busy looking for the next cow."

- *Gordon Ramsay*

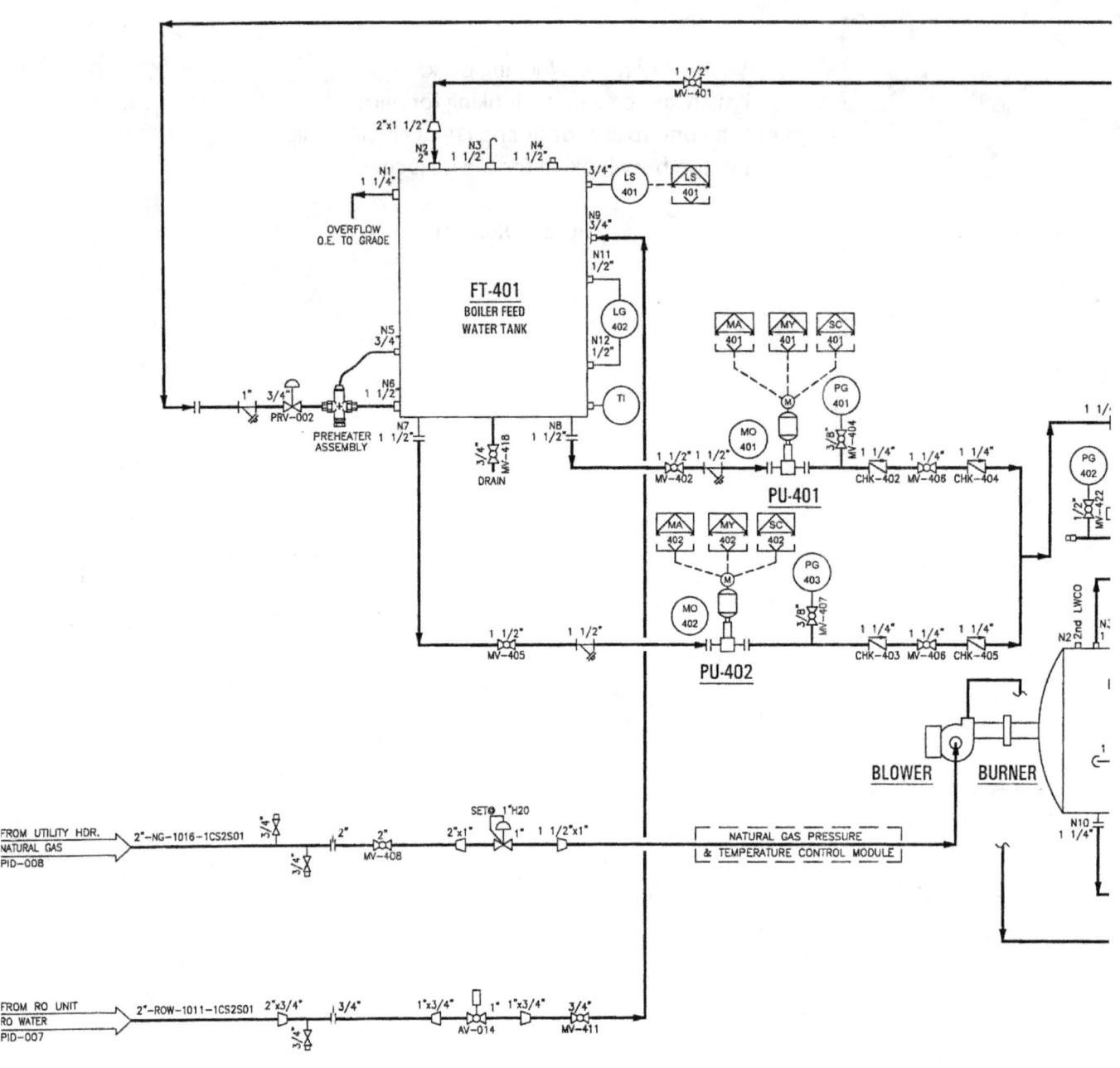

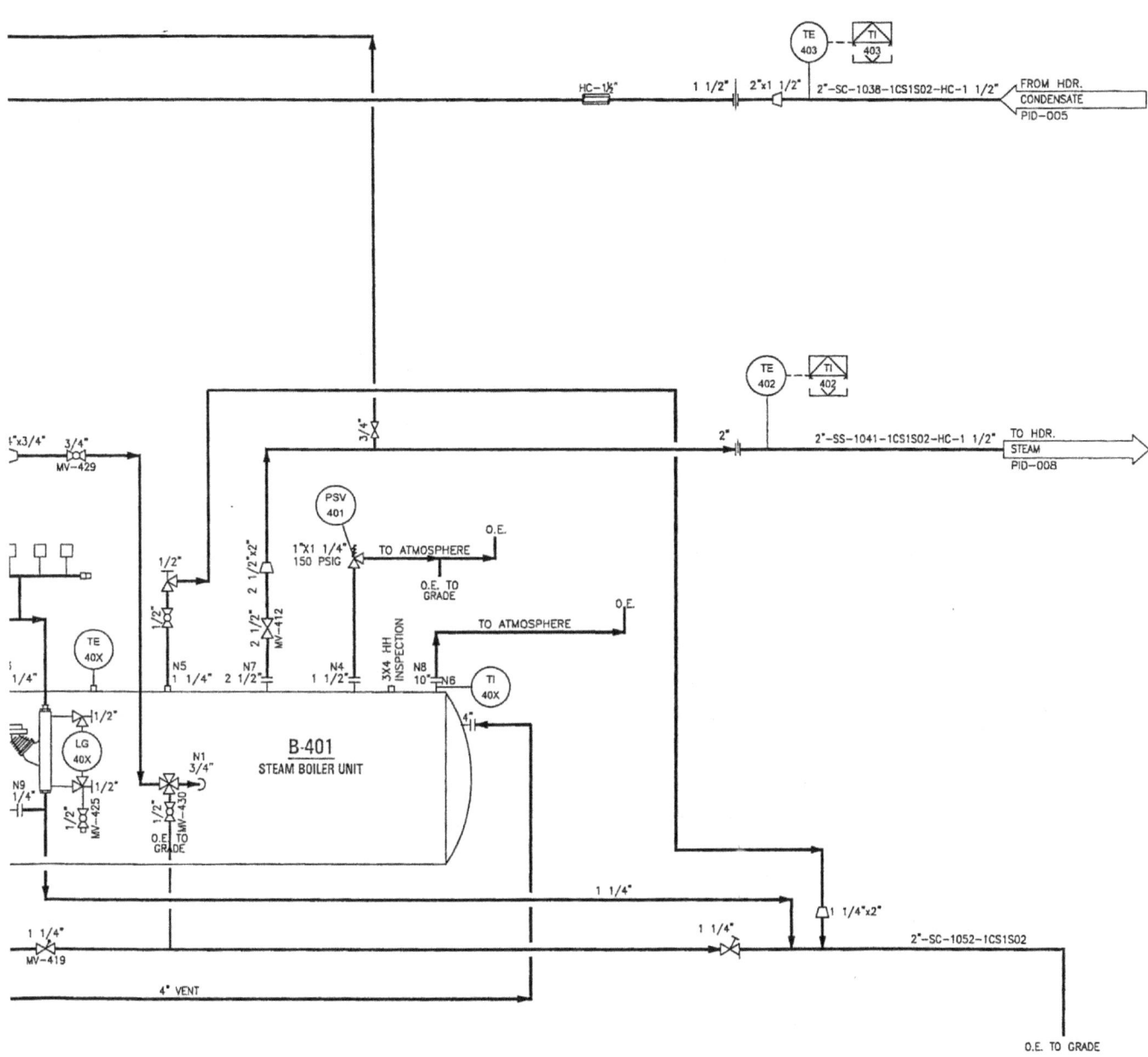

"In case you never get a second chance:

don't be afraid!

And what if you do get a second chance?

You take it!"

- *C. JoyBell C.*

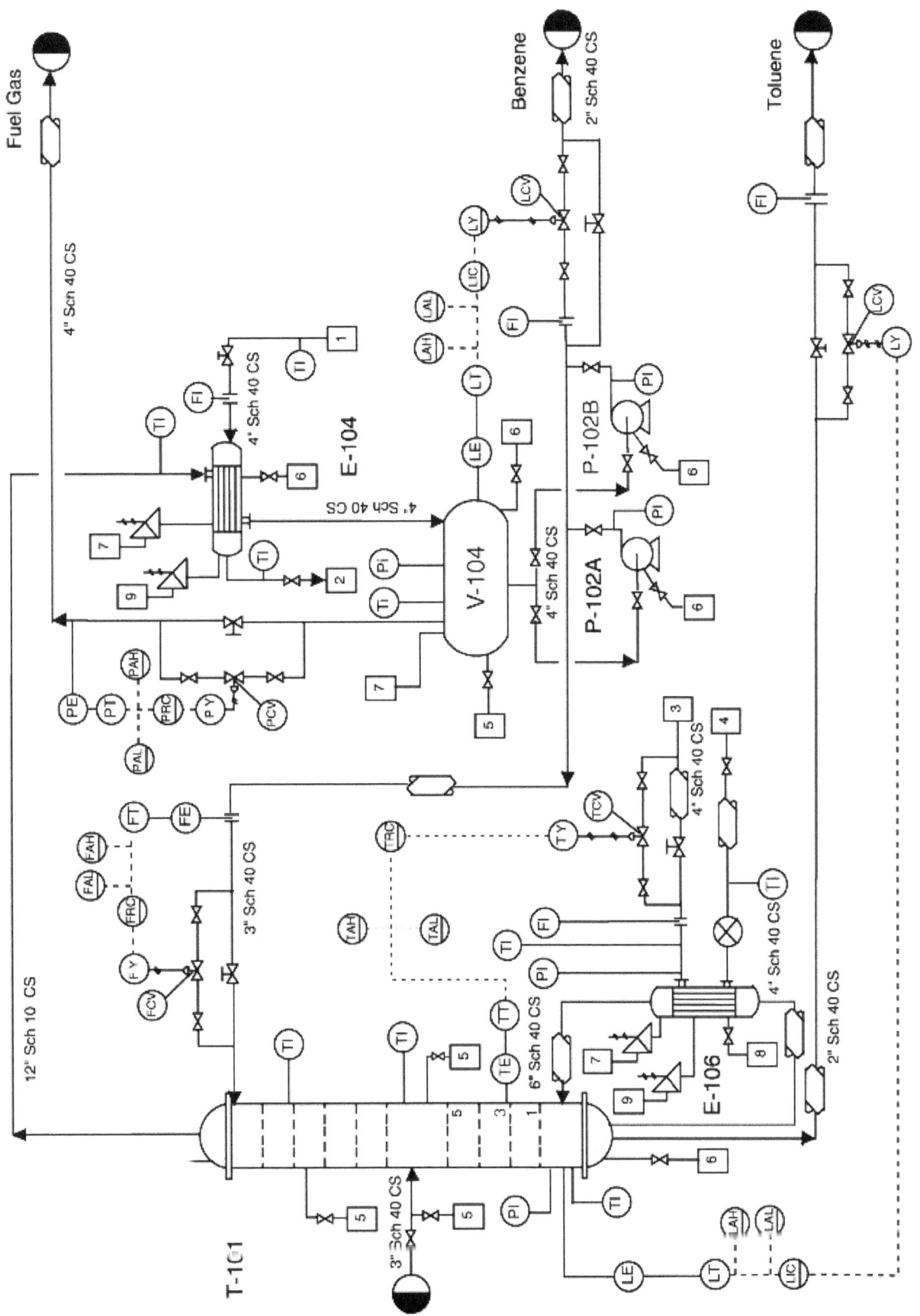

107

"Second chances do come your way.

Like trains, they arrive and depart regularly.

Recognizing the ones that matter is the trick."

- *Jill A. Davis*

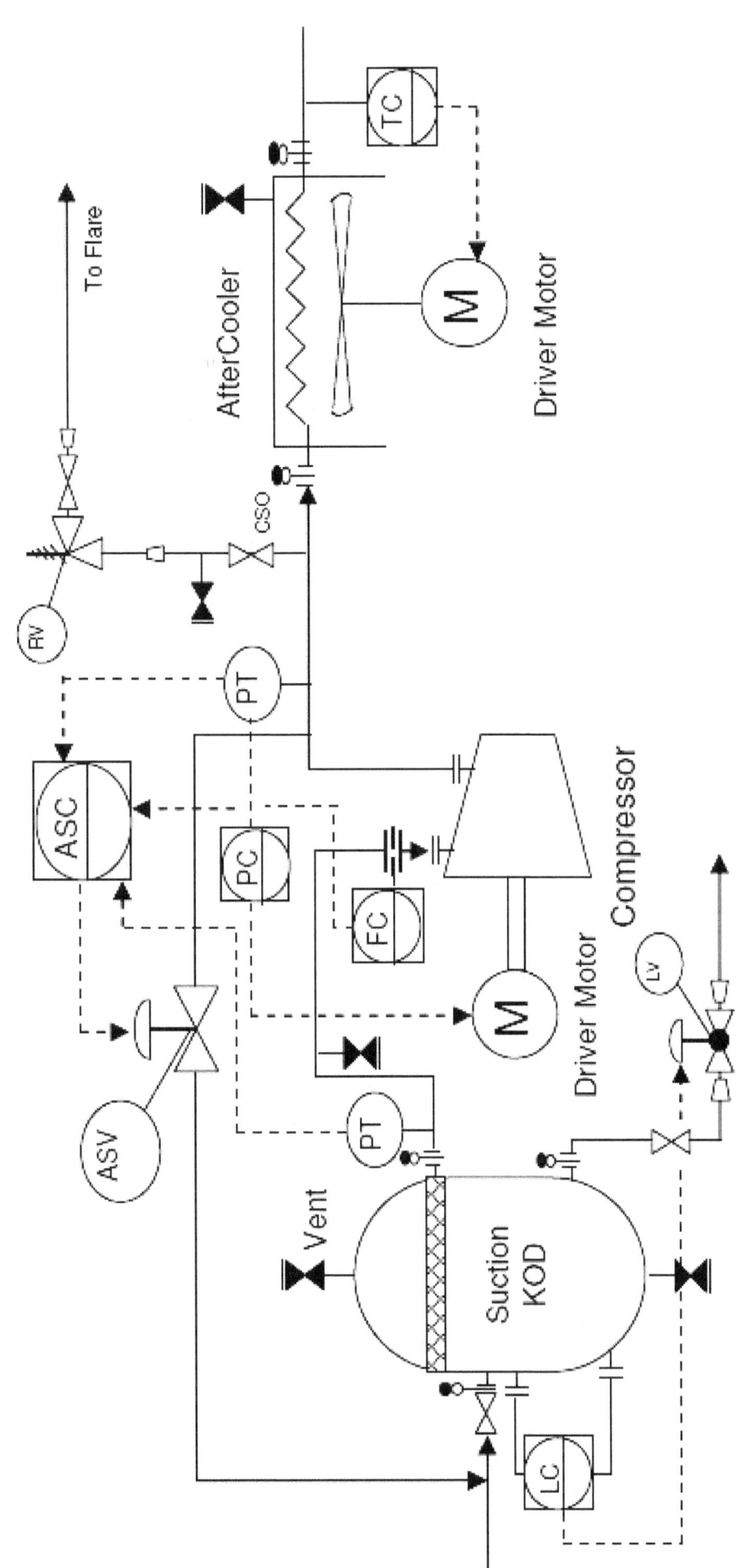

"We're given second chances every day of our life.
We don't usually take them, but they're there for the taking."

- *Andrew M. Greeley*

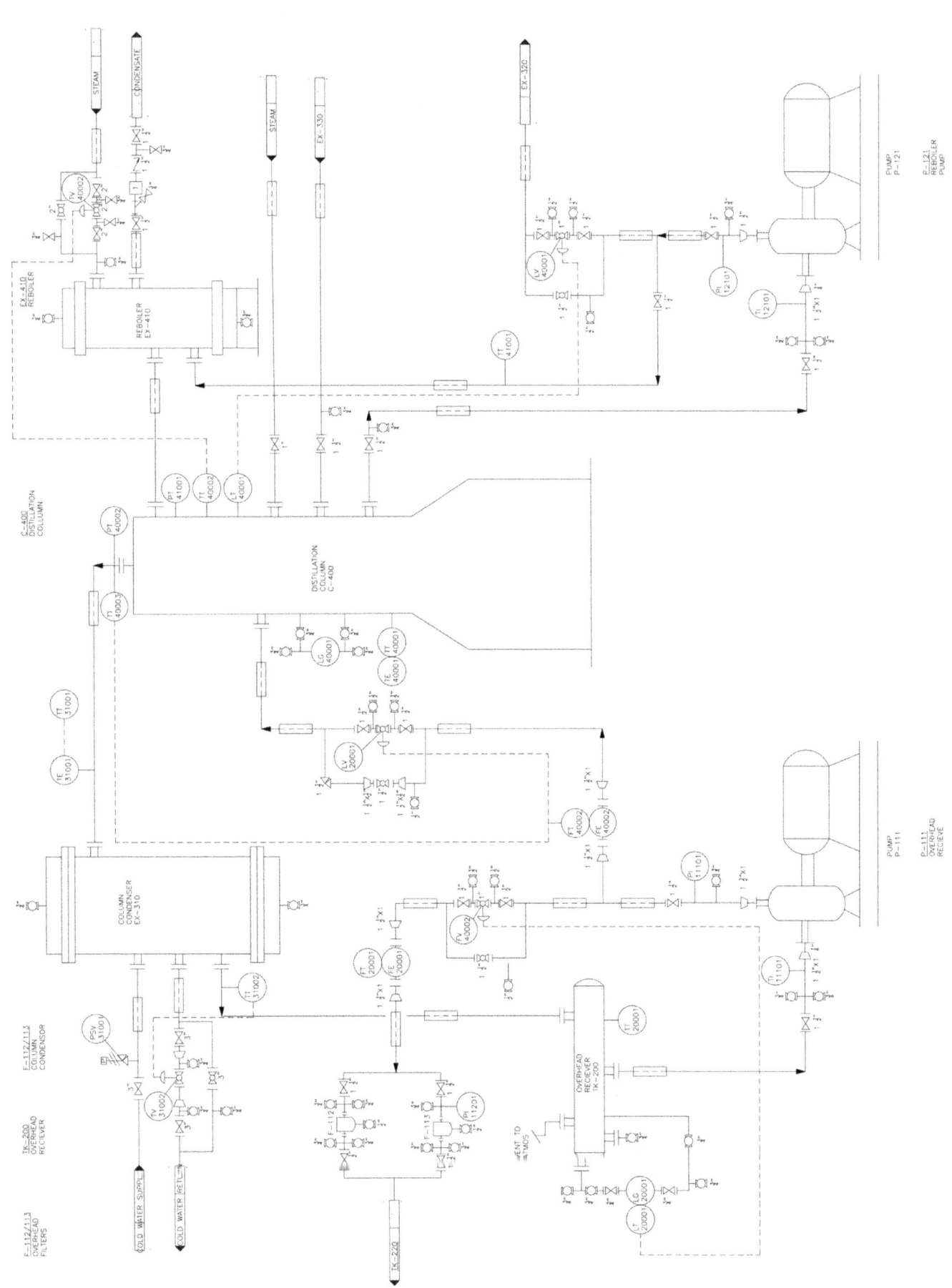

"A second chance doesn't mean anything
if you didn't learn from your first."

- *Anurag Prakash Ray*

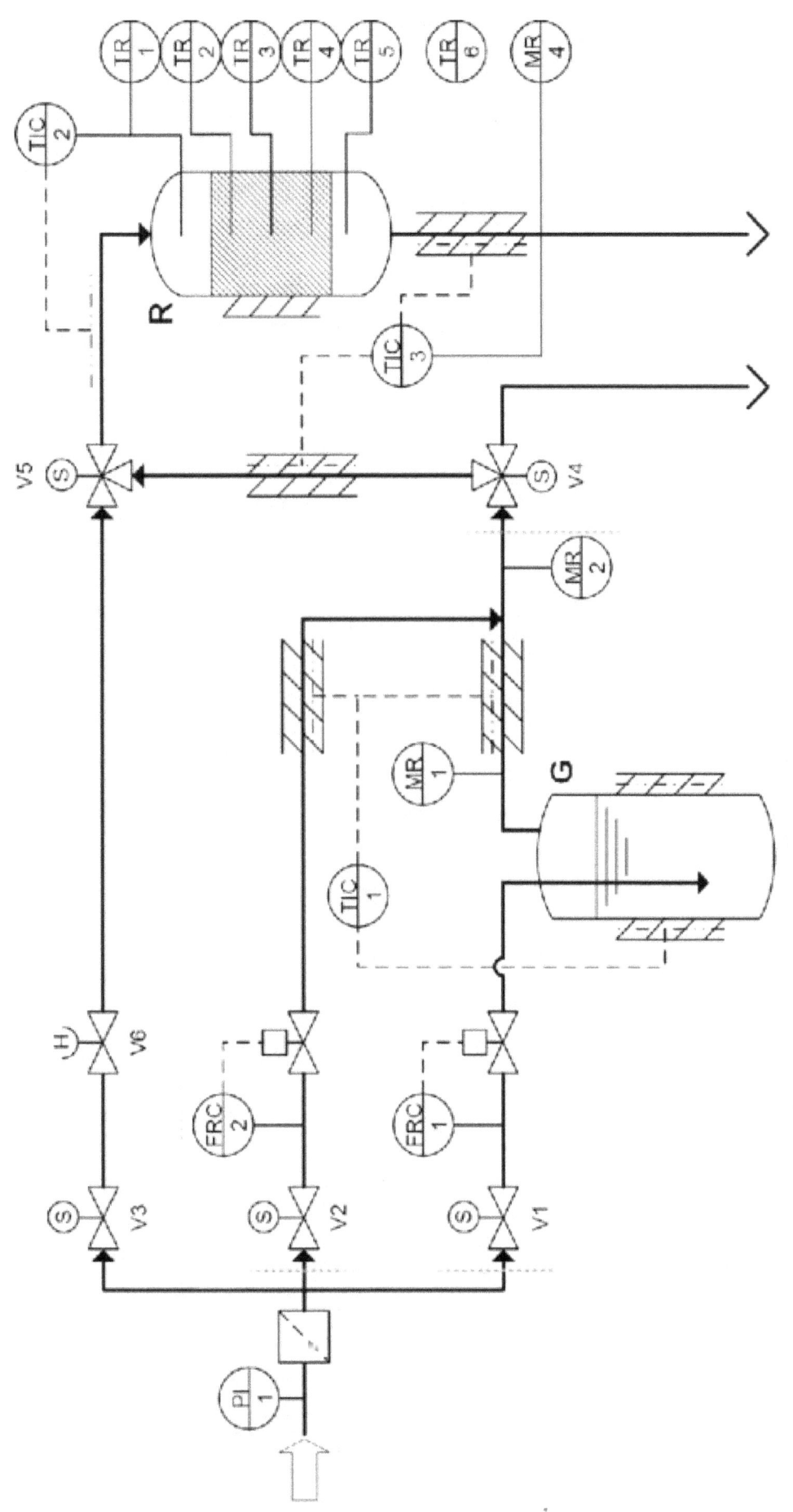

"Having a second chance
makes you want to work even harder."

- *Tia Mowry*

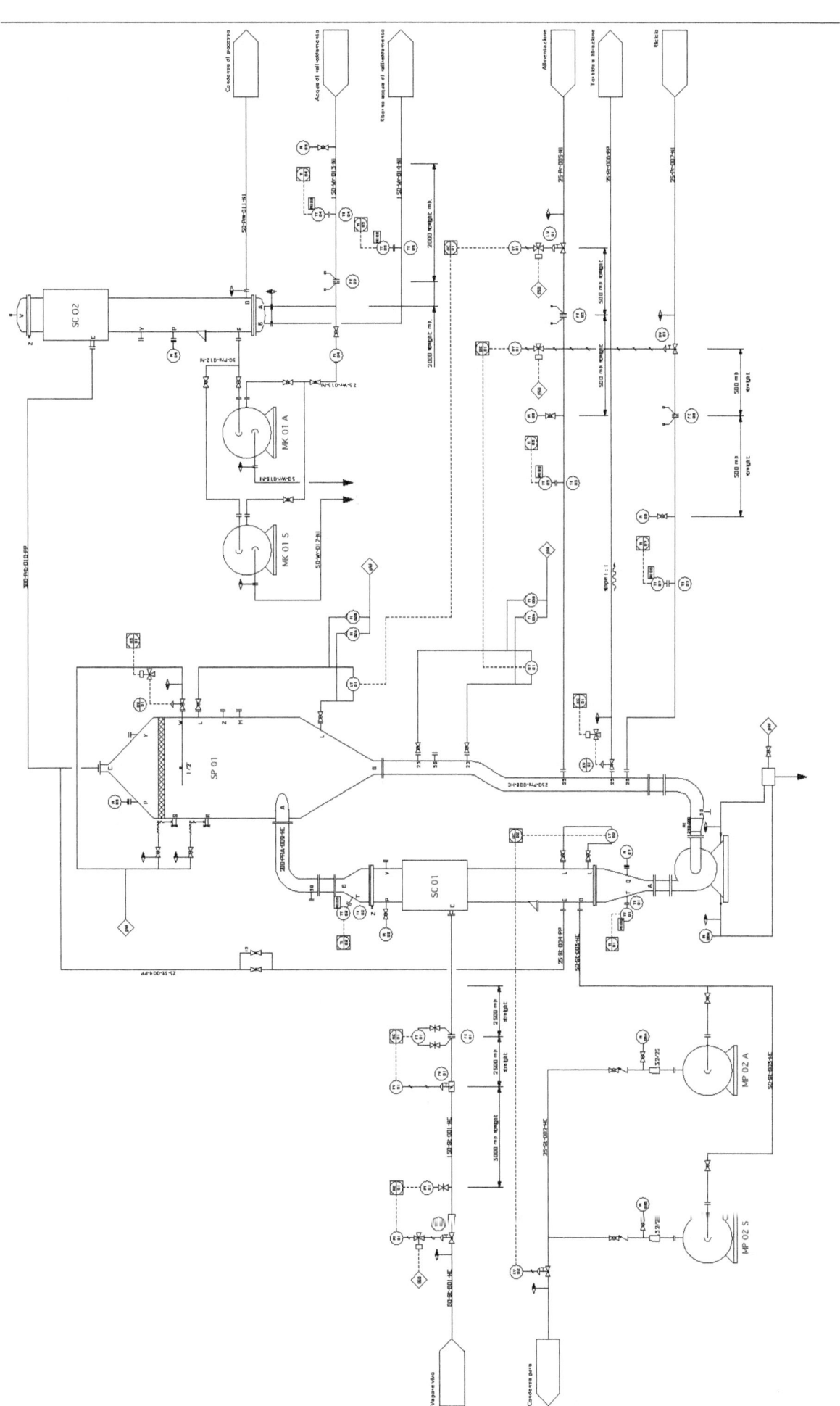

"If somebody is gracious enough

to give me a second chance,

I won't need a third."

- *Pete Rose*

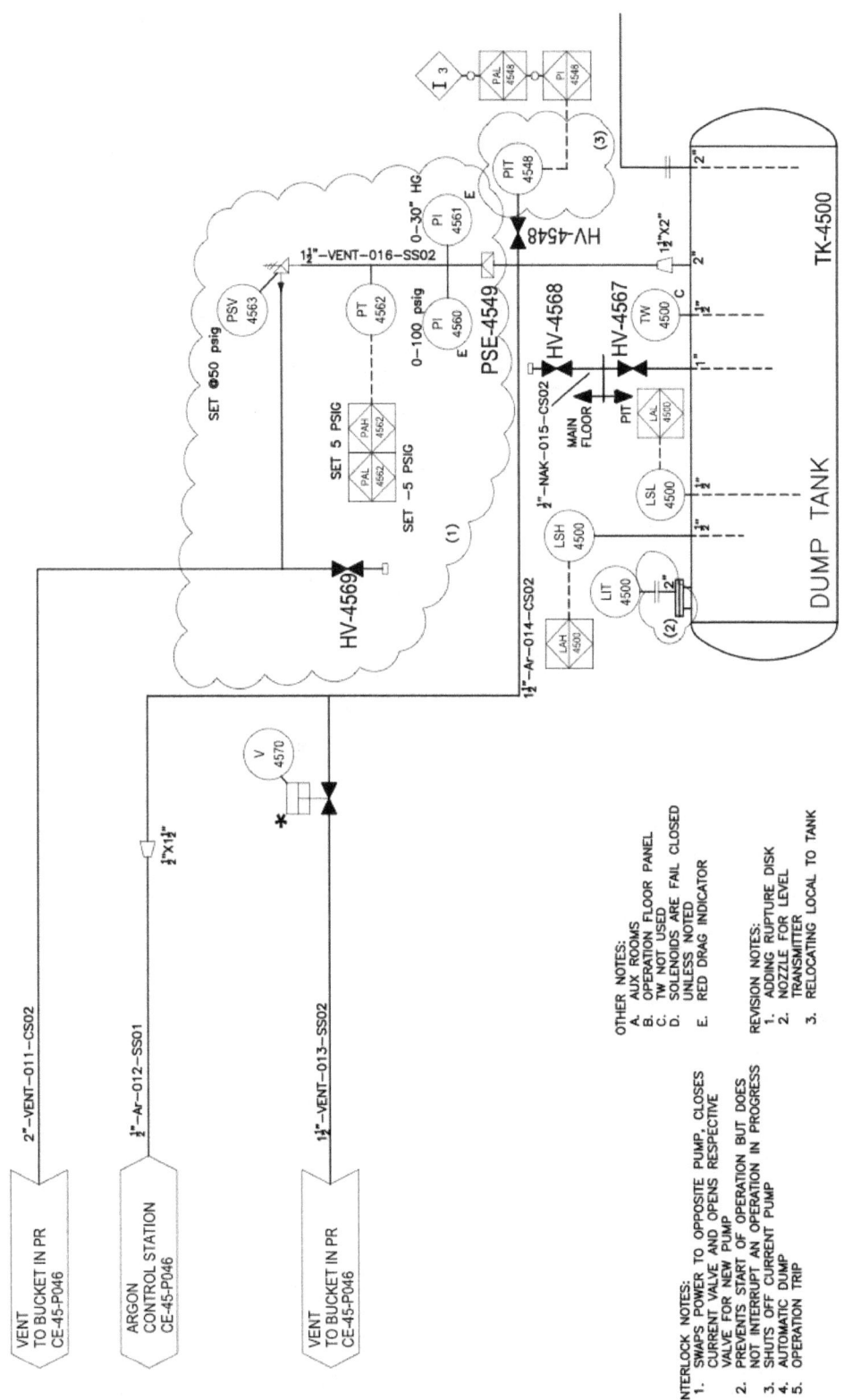

"Because this is what I believe –

that second chances are stronger than secrets.

You can let secrets go.

But a second chance?

You don't let that pass you by."

- *Daisy Whitney*

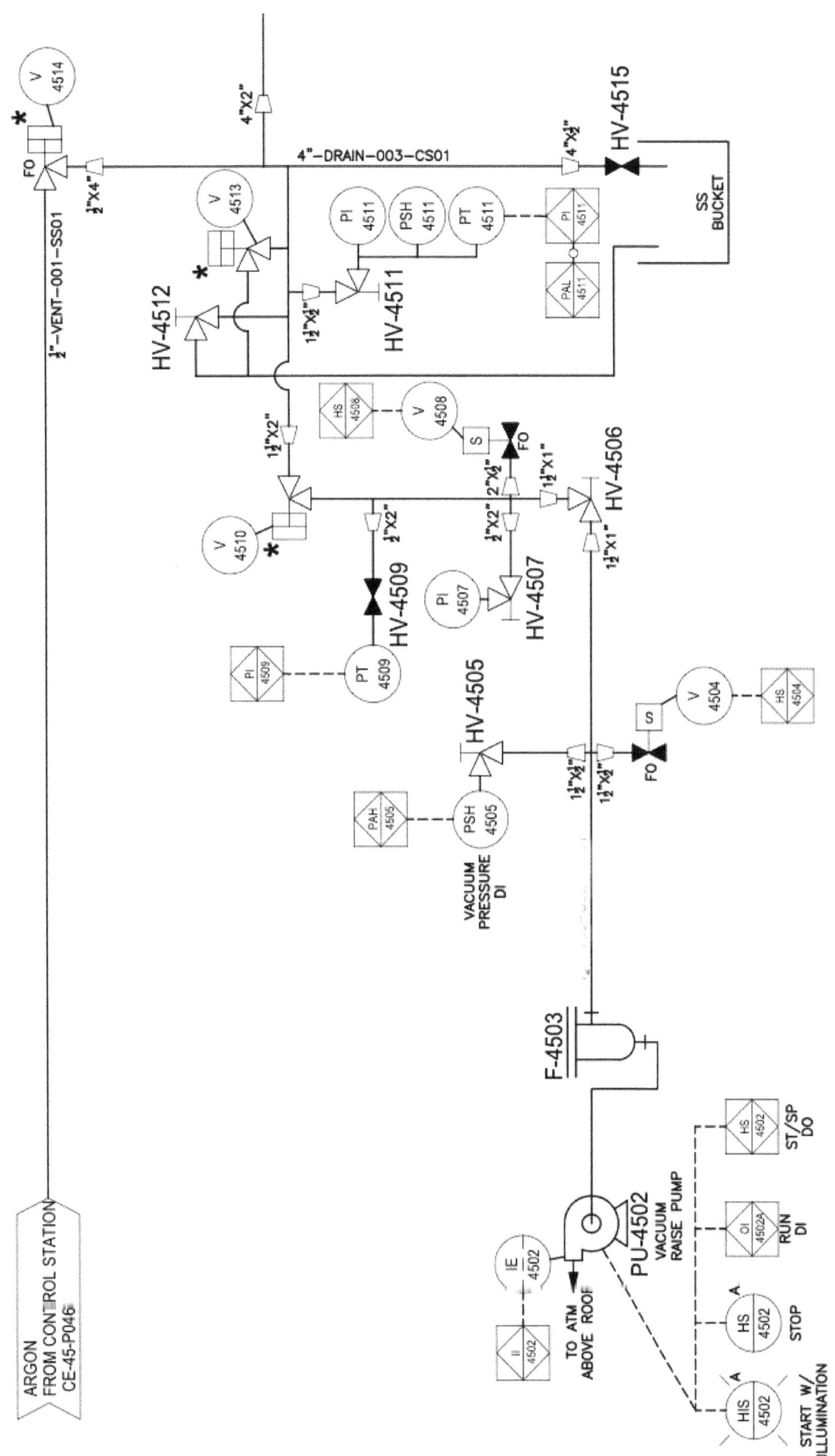

"Every moment of your life

is a second chance."

- *Rick Price*

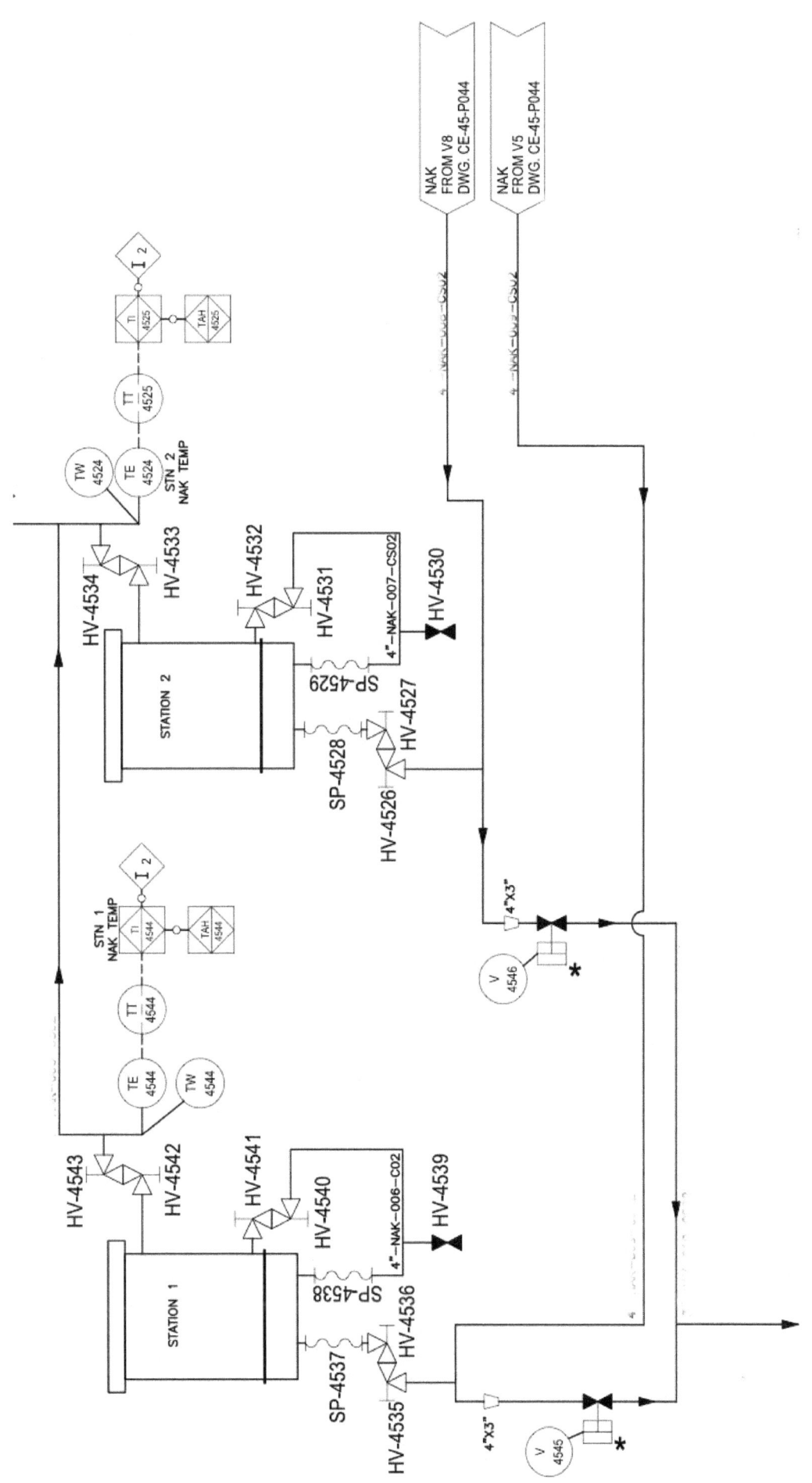

"We all deserve second chances,
but not for the same mistake."

- *Thabiso Owethu Xabanisa*

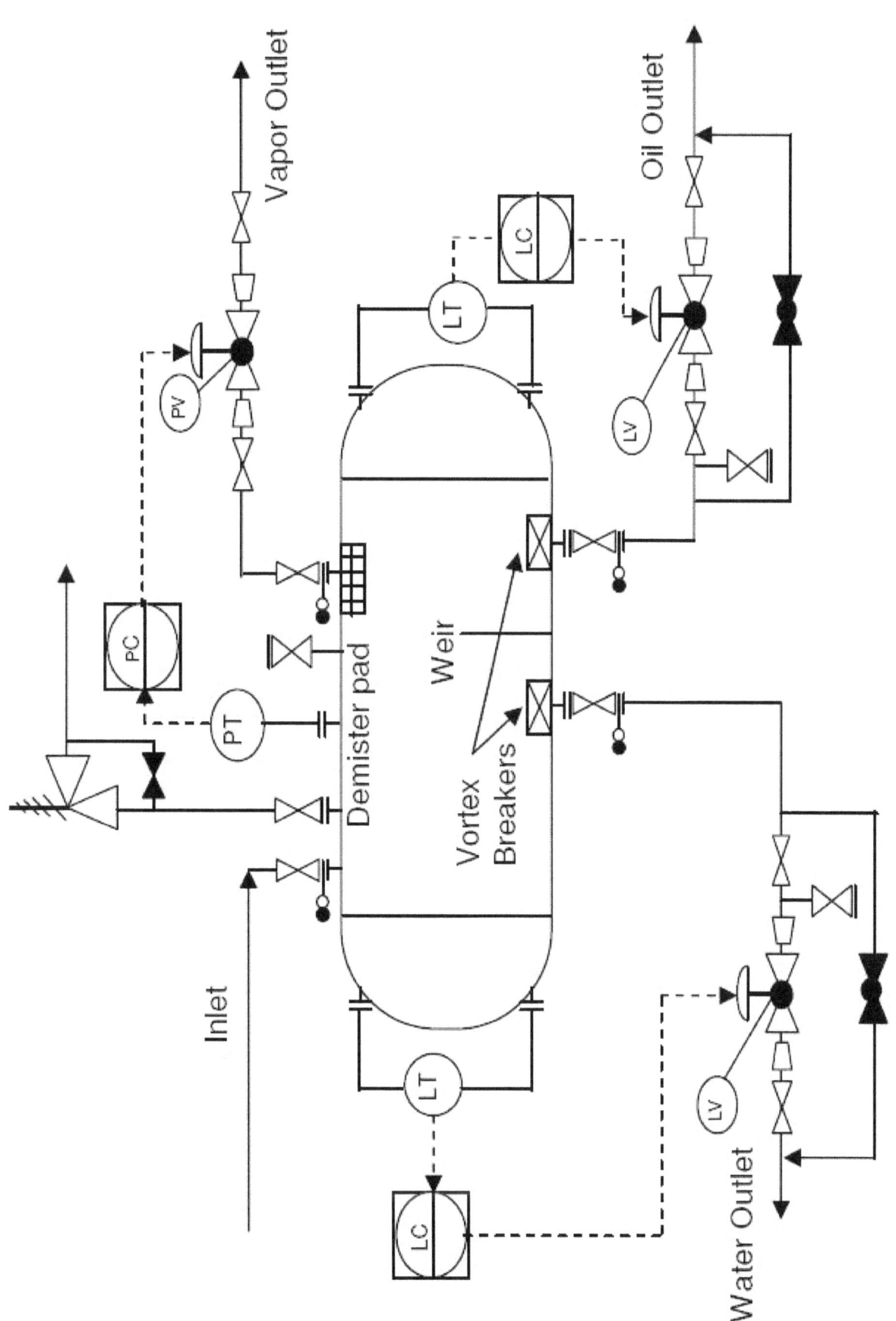

"If you ever get a second chance

in life for something,

you've got to go all the way."

- *Lance Armstrong*

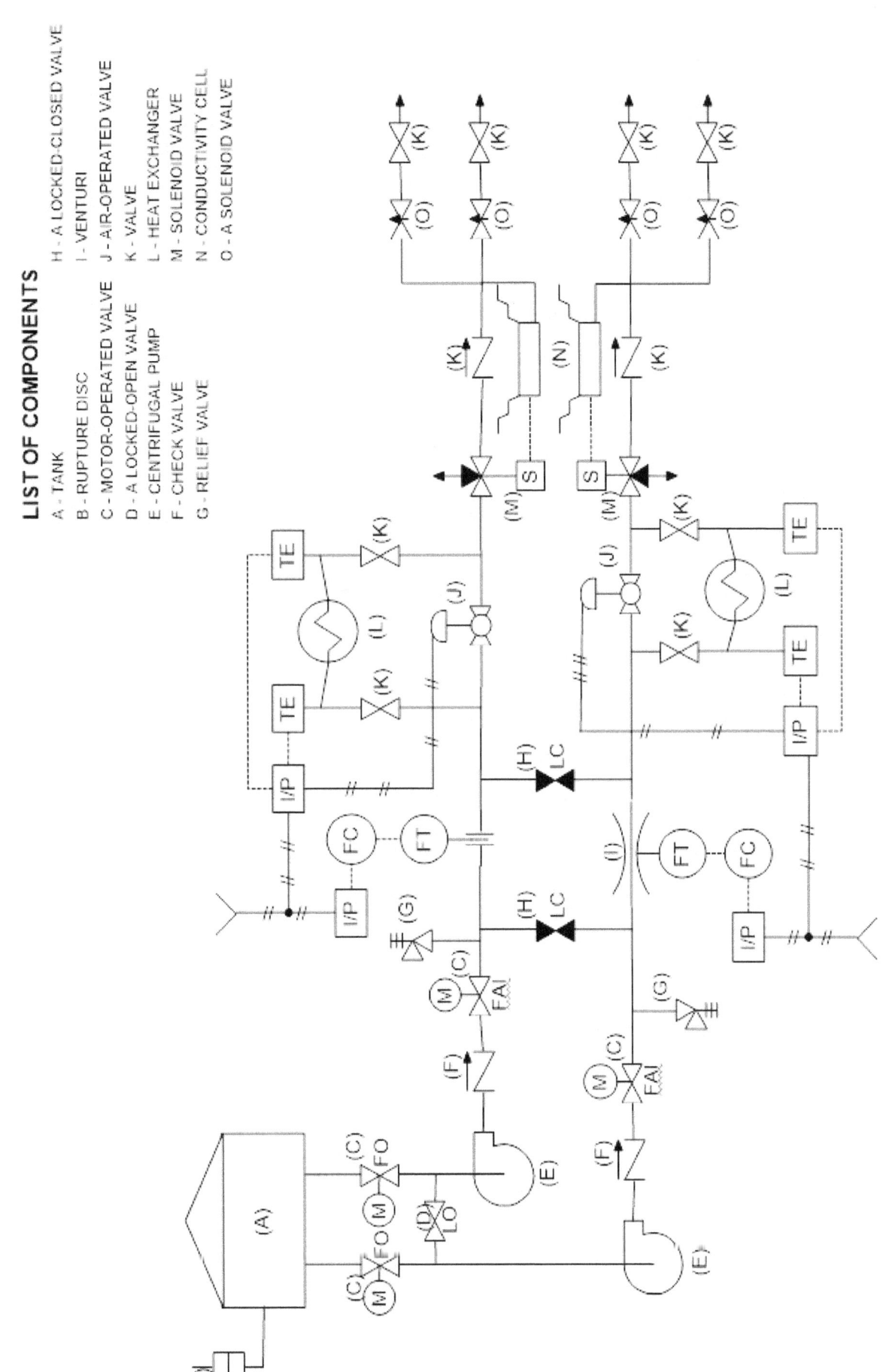

"A man deserves a second chance,
but keep an eye on him."

- *John Wayne*

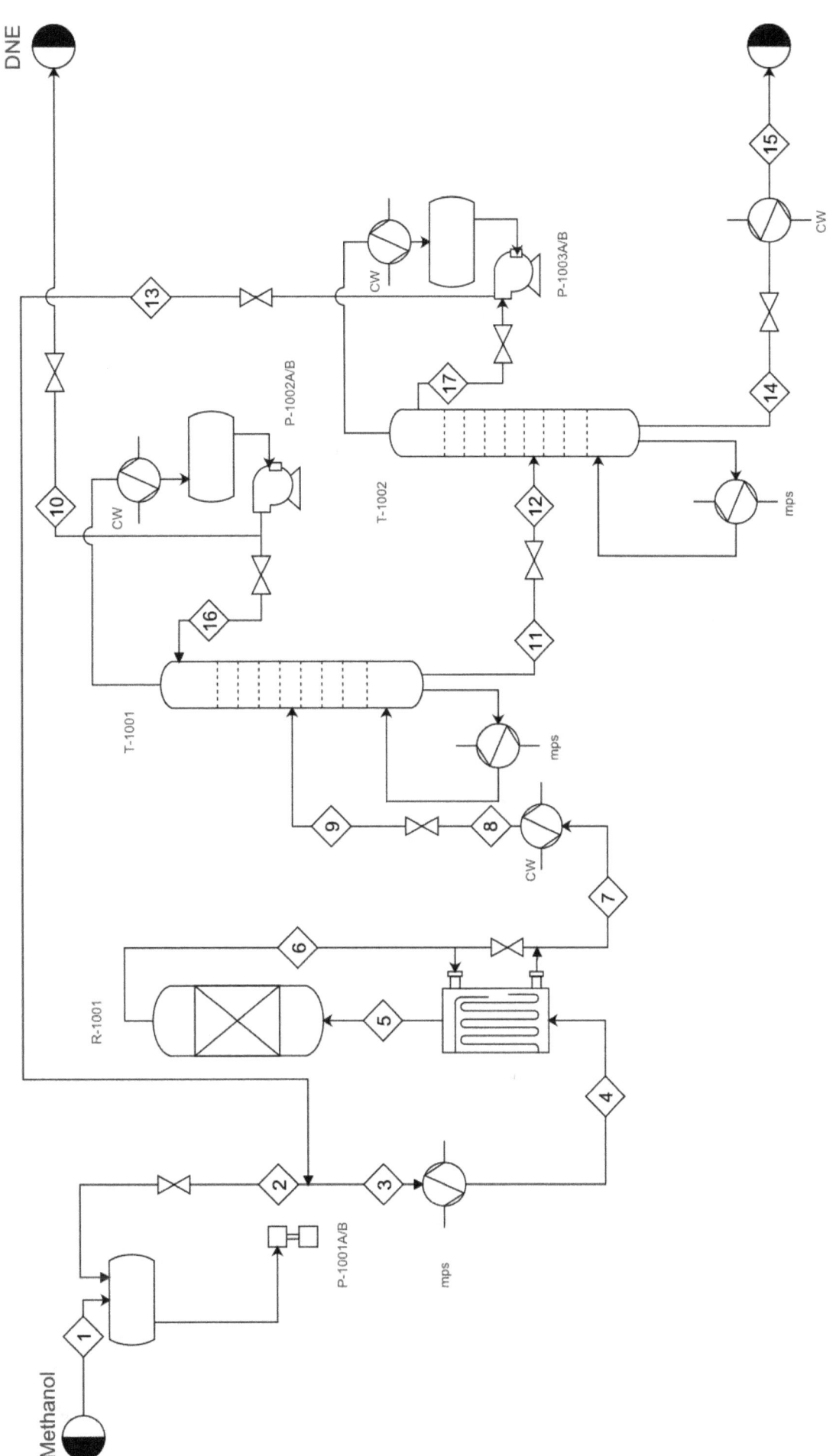

"If you are still breathing,
you have a second chance."

- Oprah Winfrey

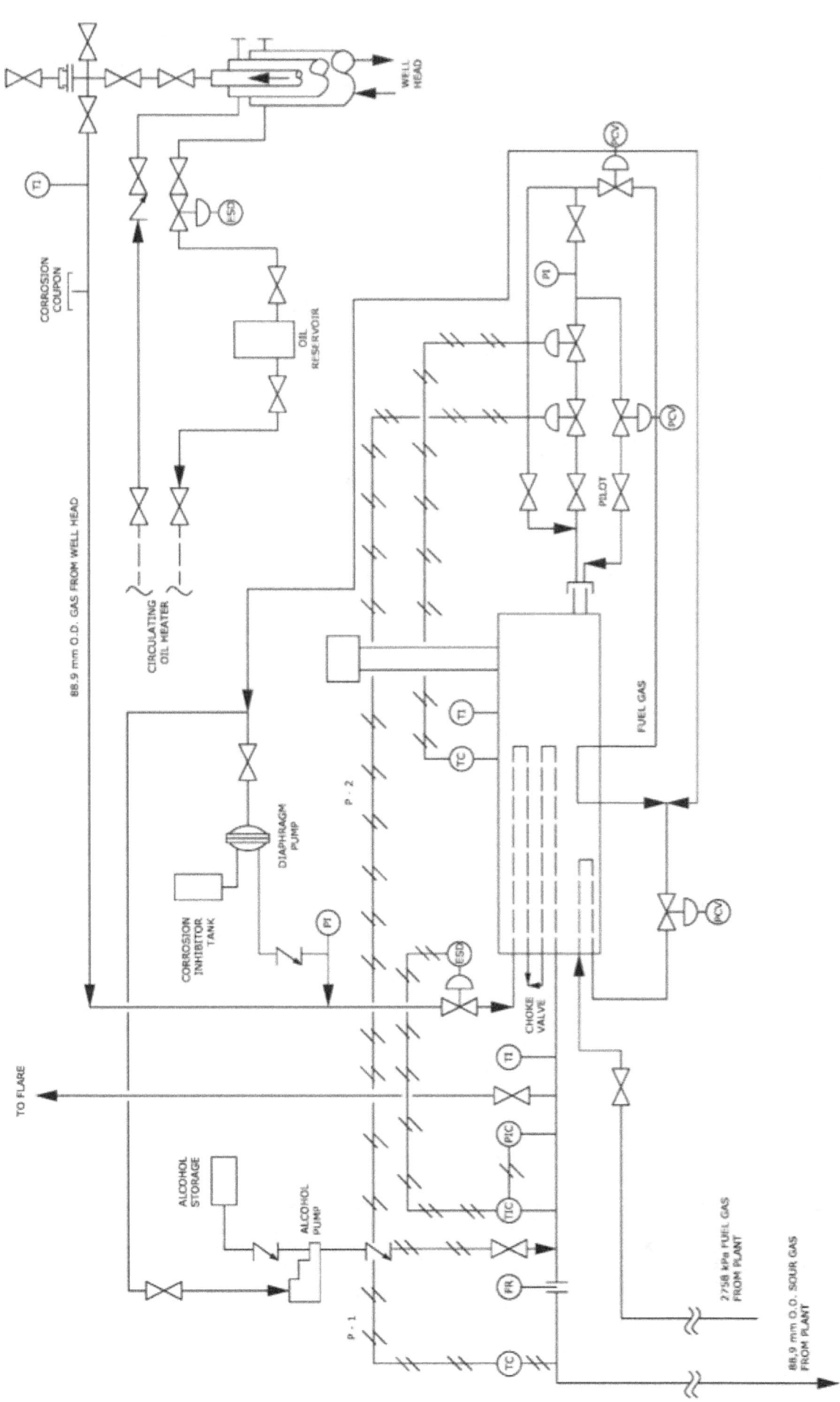

"I believe in second chances,

but I don't believe

in third or fourth chances."

- *Selena Gomez*

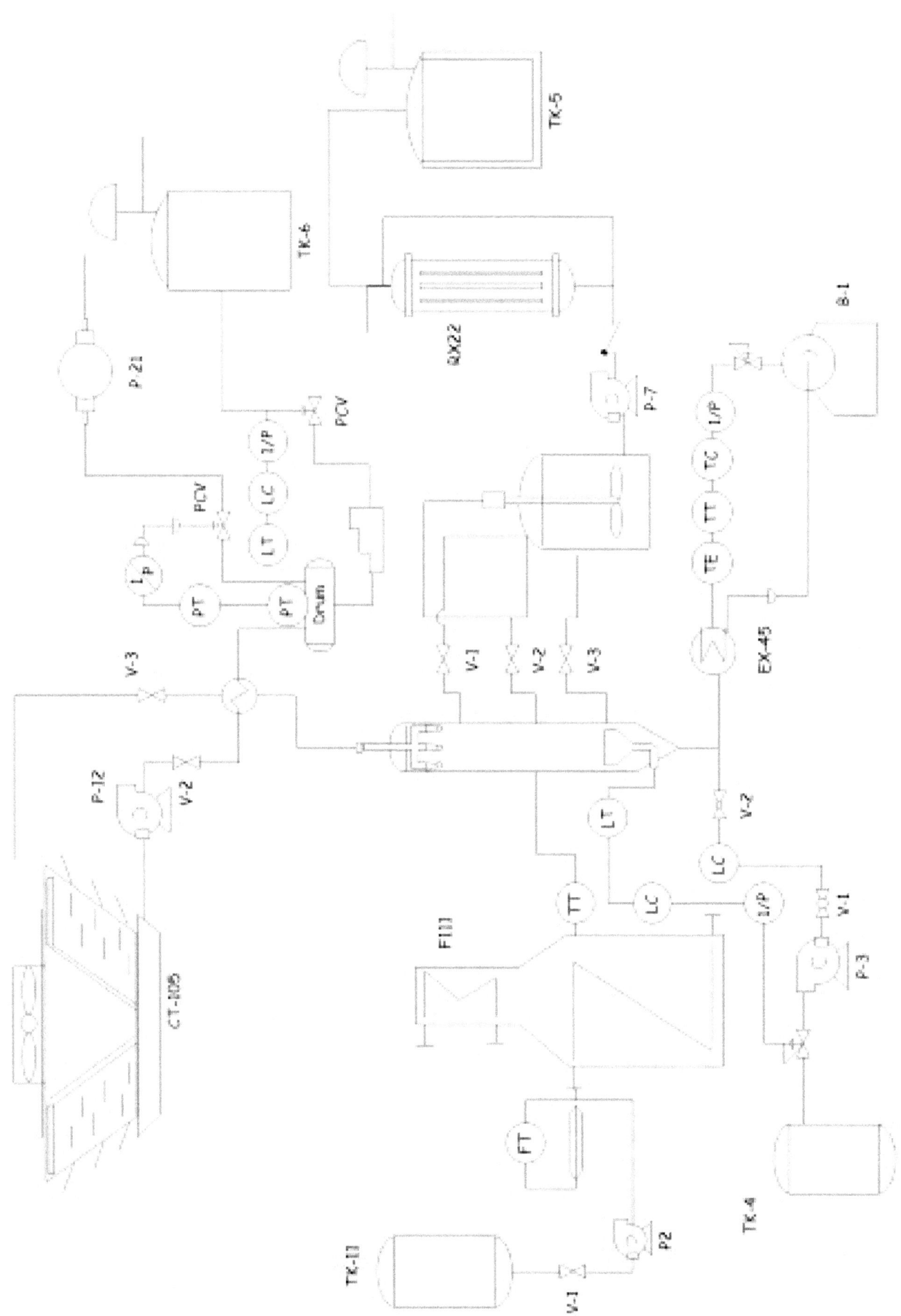

"Every second a seeker can start over,

For his life's mistakes are initial drafts

And not the final version."

- *Sri Chinmoy*

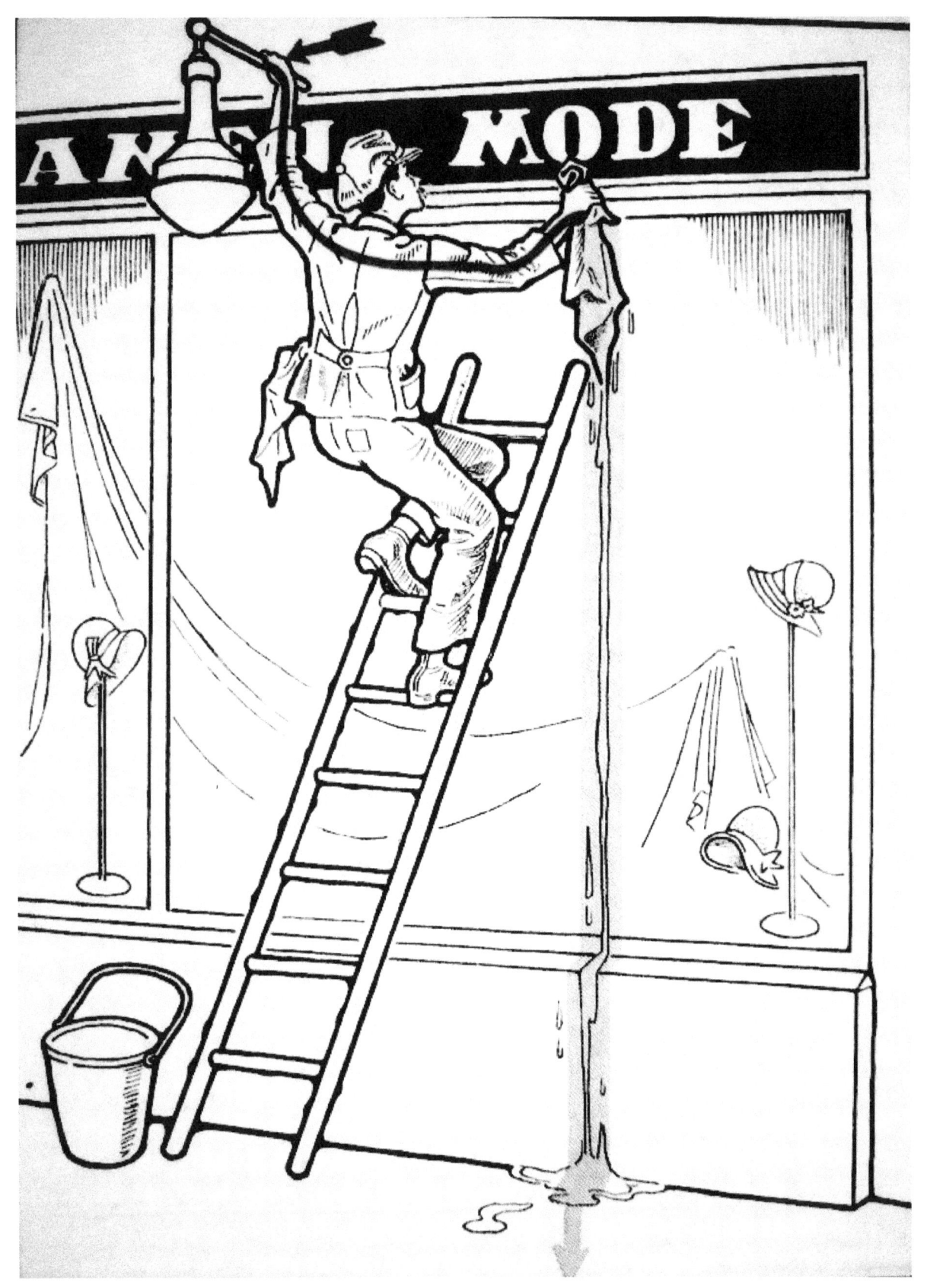

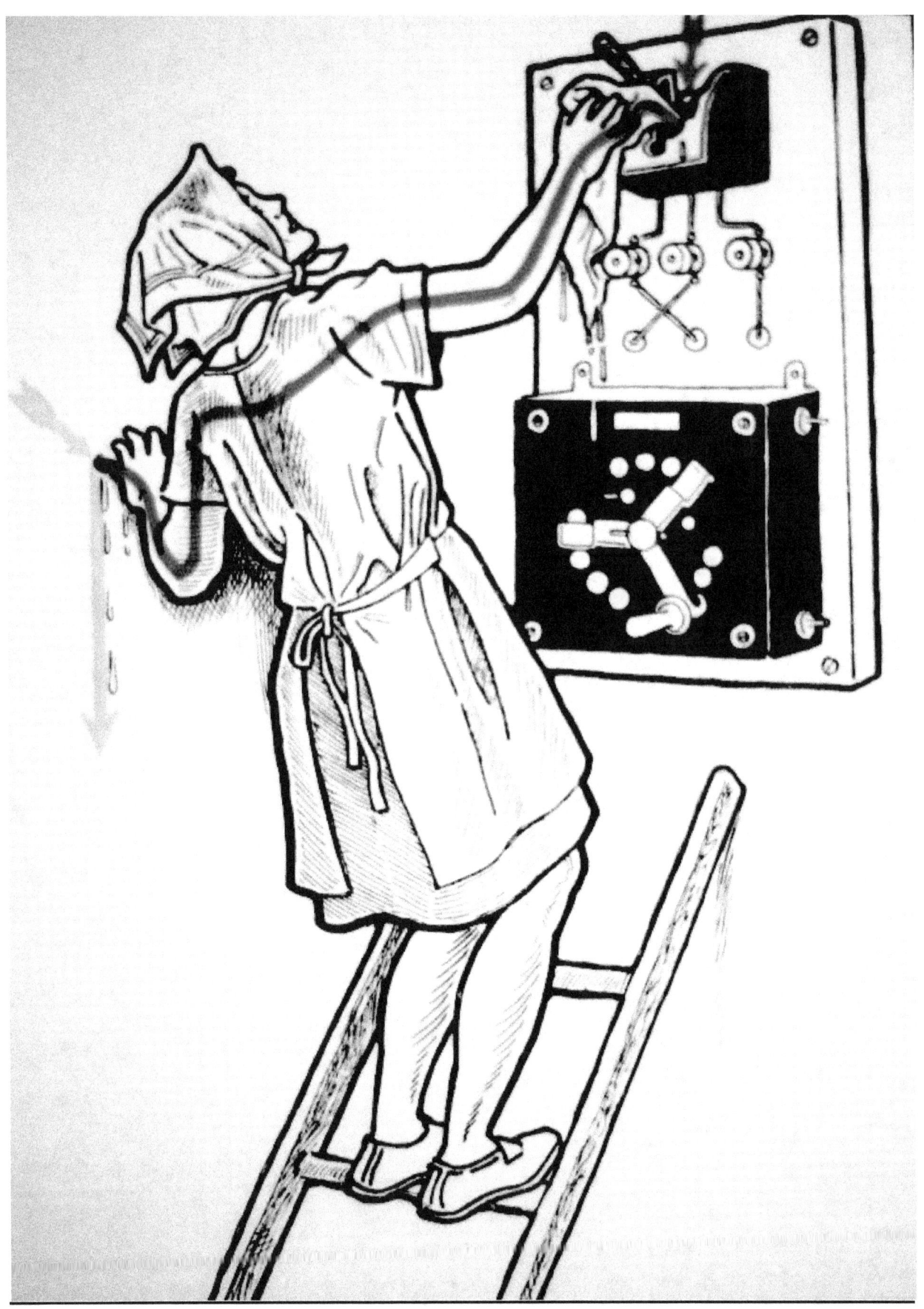

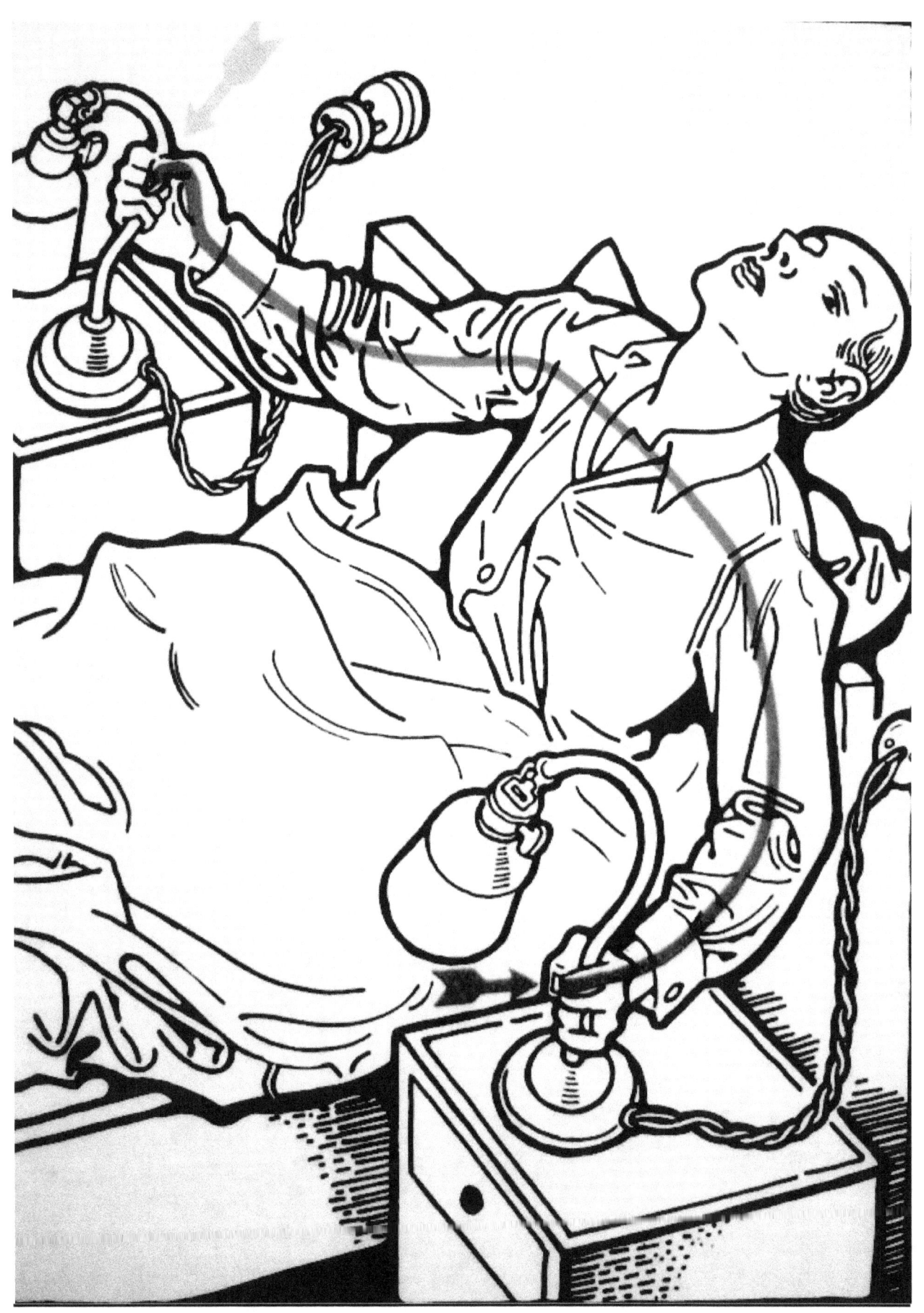

159

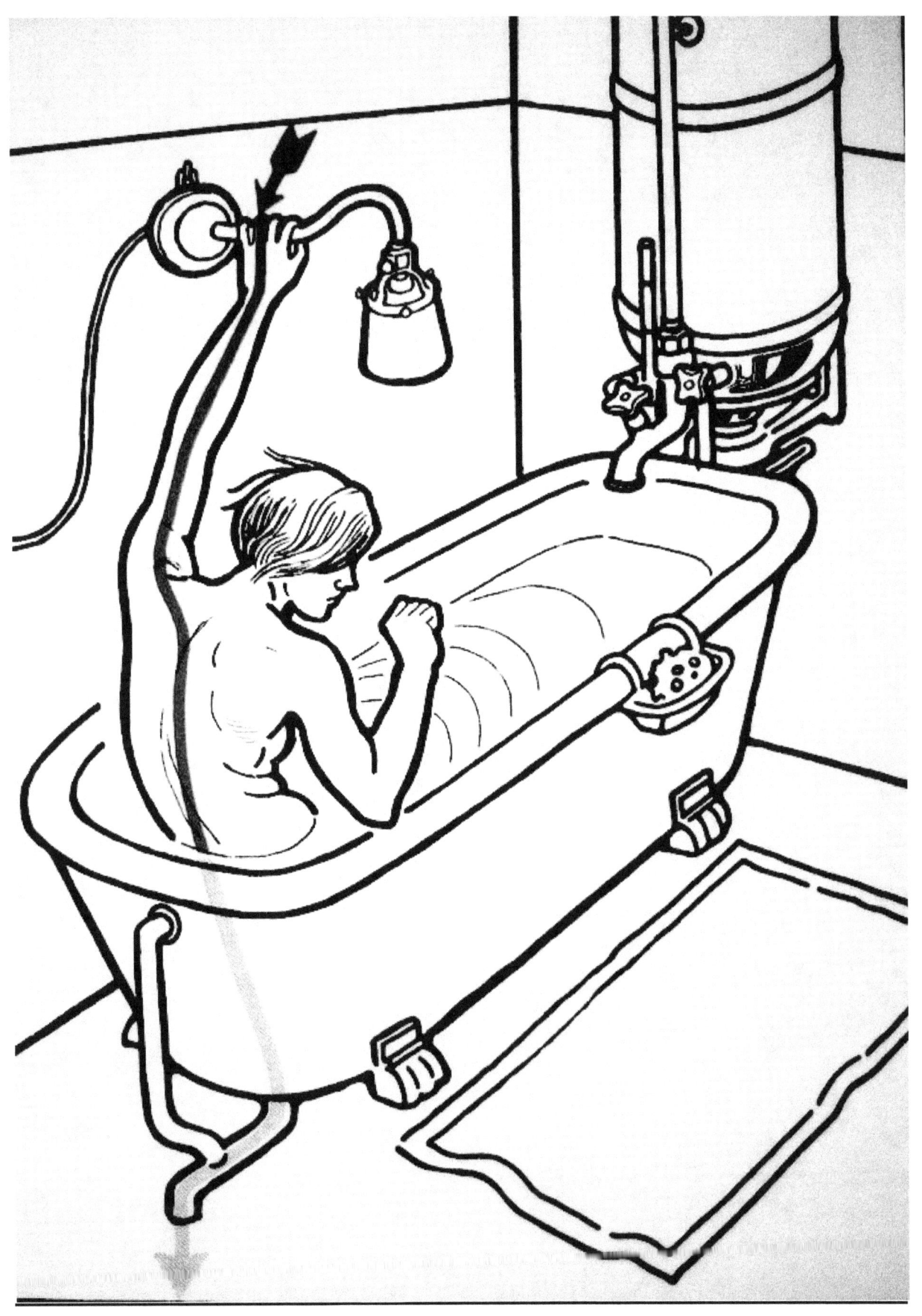

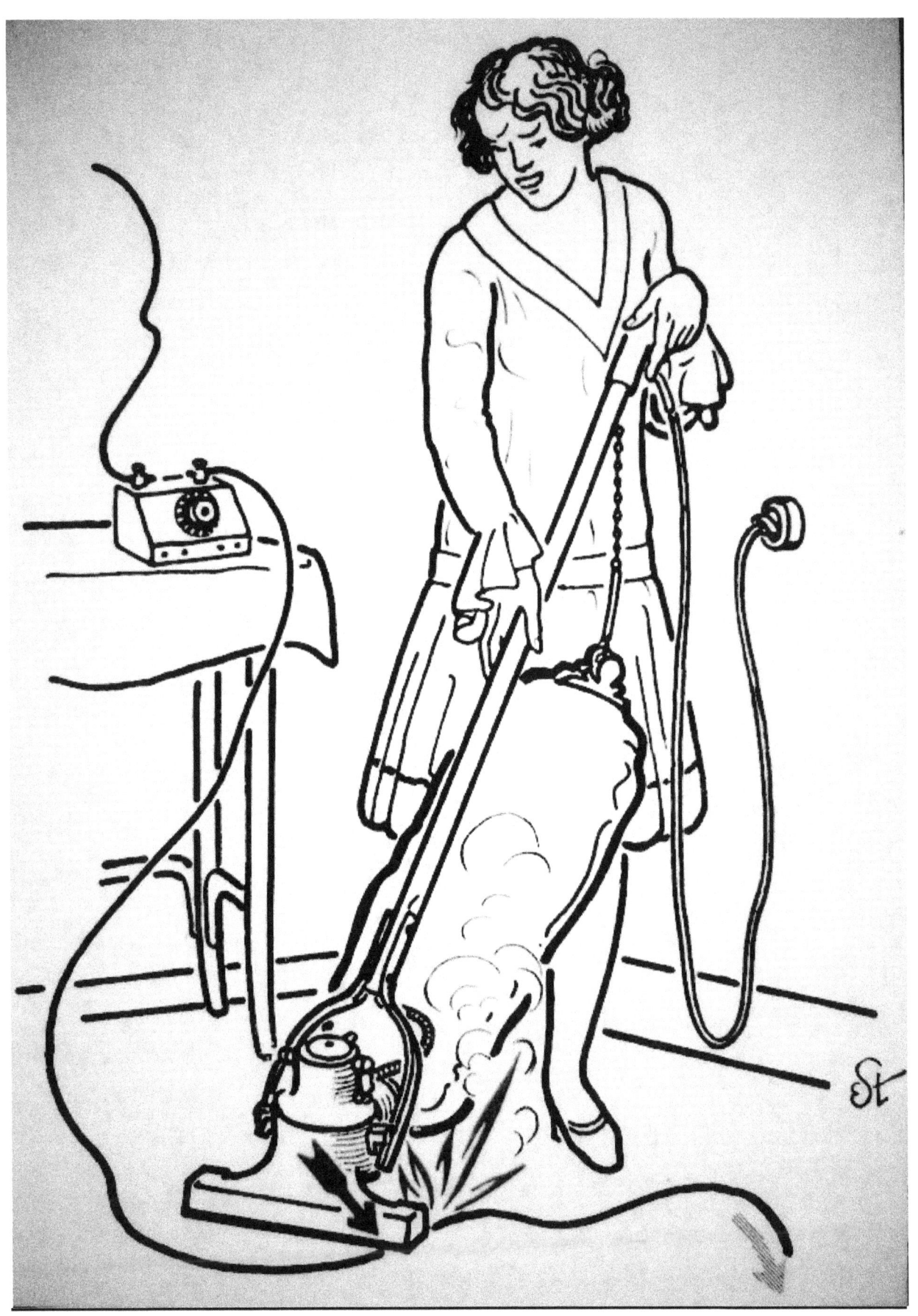

183

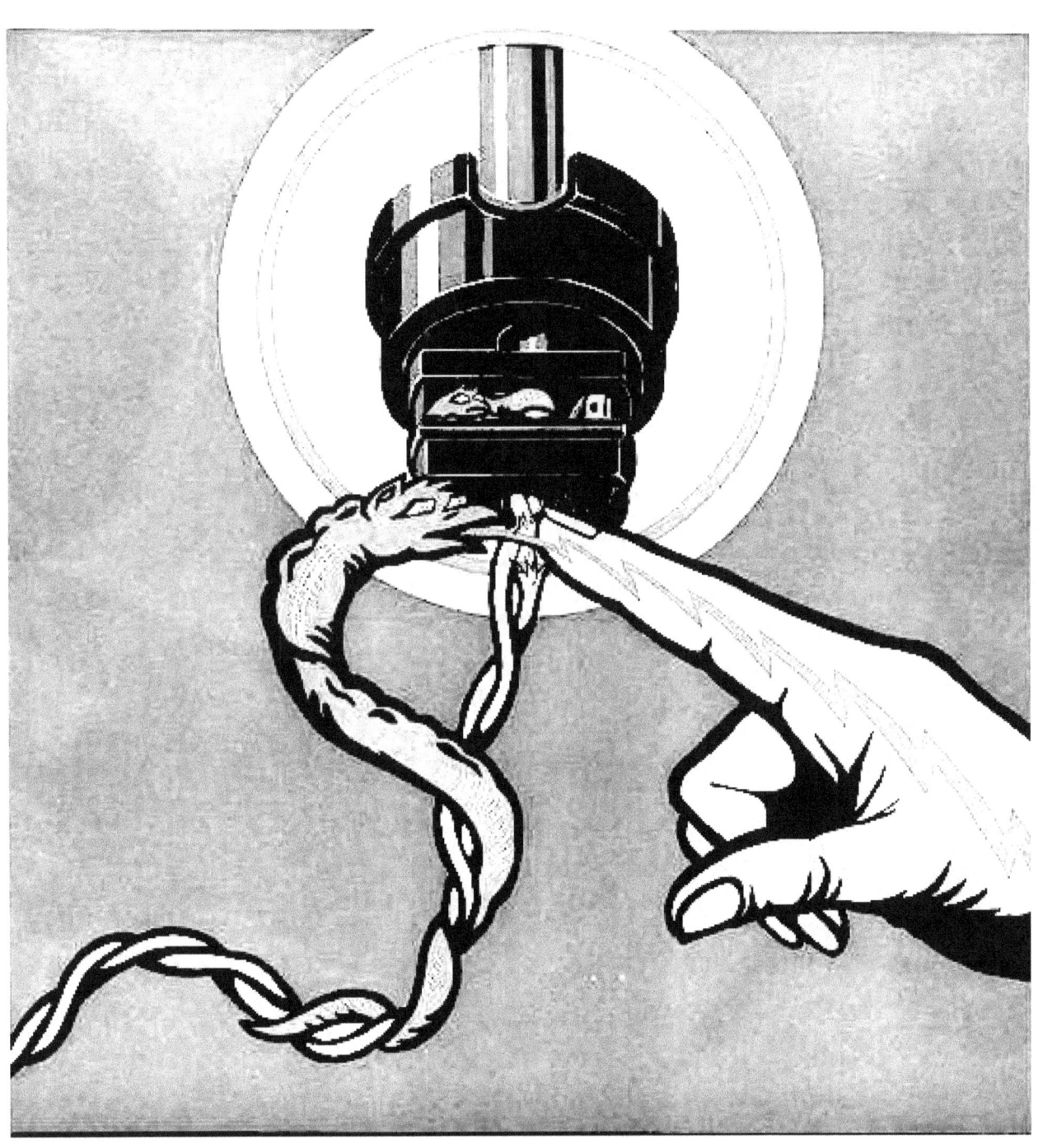

WE TRAIN TECHNICIANS

PAC Works
219 Defee St.
Baytown, TX

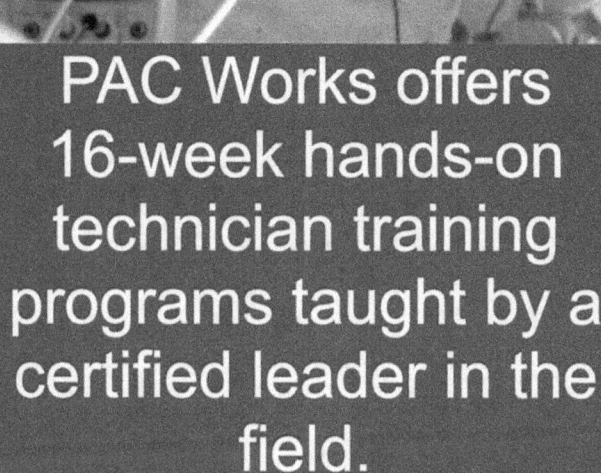

PAC Works offers 16-week hands-on technician training programs taught by a certified leader in the field.

Call Us | (832) 926-4717

Ezekiel Caudill was born in Joliet, Illinois.
He joined the United States Army after high school
and studied Process Instrumentation & Electrical Design at Lee College.
He is a proud resident of Baytown, Texas.

Fox Howell was born in Hinesville, Georgia.
He joined the United States Navy after high school
and studied at Lee College in Baytown, Texas.

PAC Technical Publication Group is an alliance of technical experts
who work together to improve the level of technical training available to students.
The group has a combined total of more than 141 years of technical experience
between them, and is based in Baytown, Texas.

This book is published by PAC Technical Publication Group.

Contact PAC Technical Publication Group at pactechpubgroup@gmail.com

www.ingramcontent.com/pod-product-compliance
Lightning Source LLC
Chambersburg PA
CBHW080543220526
45466CB00010B/3015